GRADE 4

SCIENCE
A CLOSER LOOK

BUILDING SKILLS

Reading and Writing Workbook

McGraw Hill Education

Contents

Contents

Contents

Contents

Contents

PHYSICAL SCIENCE

Contents

Dragons of the Sea
by Elizabeth Schleichert

Read the Unit Literature feature in your textbook.

Write About It

Response to Literature The name "leafy seadragon" sounds almost like a plant. Is a leafy seadragon a plant or an animal? How can you tell? Write an essay to compare and contrast plants and animals.

A leafy seadragon is a plant. It is a plant becaus it has leaf and seed that why I think it is a plant. Animals and plants are both leving thing. Plant grow in the durt and animals grow form eggs or more things.

Name _____ Date _____

Kingdoms of Life

Complete the concept map about the classification of living things. Some examples have been done for you.

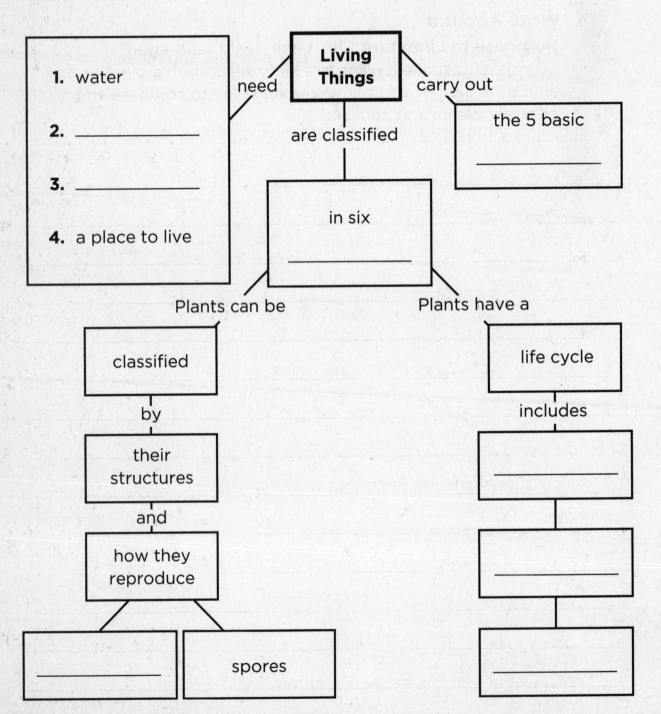

1. water
2. _____
3. _____
4. a place to live

need → **Living Things** ← carry out

the 5 basic _____

are classified

in six _____

Plants can be

Plants have a

classified

by

their structures

and

how they reproduce

_____   spores

life cycle

includes

_____

_____

_____

Cells

Use your textbook to help you fill in the blanks.

What are living things?

1. People, _____ , and _____ are living things.

2. Living things need water, food, a place to live, and most

   need _____ to survive.

3. All living things perform five basic jobs, or life functions.

   a. They use _____ for energy.

   b. They _____ and develop.

   c. They _____ more of their kind.

   d. They respond to their _____ .

   e. They get rid of _____ .

4. All living things, also called organisms, are made

   of _____ .

How do plant and animal cells compare?

5. All cells have smaller parts that work to keep the

   cell _____ .

6. Plant cells contain _____ , a substance

   that plants use to capture the _____
   energy to make food.

7. Animals cannot make their own _____
   because they do not have chlorophyll.

Name _____ Date _____

How are cells grouped?

8. Cells are grouped by the _____ they do.

9. A group of similar cells that carries out a certain job is

   called a _____ .

10. Tissues in a group are called an _____ .

11. You have many organs that work together in an organ

    _____ .

How can you see cells?

12. A microscope works like a hand lens by making

    something _____ look much

    _____ .

Critical Thinking

13. Which do you think would be more harmful to an organism:
    a damaged cell or a damaged organ?

    _____

    _____

    _____

    _____

    _____

    _____

    _____

Cells

Use the clues below to complete the word puzzle.

Across

2. living thing

4. young organisms of parents

6. rigid outer coverings of plant cells

7. similar cells working together

Down

1. 5 basic jobs of living things

2. organs working together

3. tissues working together

4. gas in the air

5. smallest part of living thing

Cells

Fill in the blanks.

food	life functions	offspring	oxygen
grow	living	organisms	small

Everything in the world can be placed into one of two

groups. There are _____ things and

nonliving things. All living things need water, food, a place to

live, and most need _____ . Also, all living

things carry out five _____ . First, they need

_____ for energy. Second, living things

_____ and develop. Third, they respond

to the environment, and fourth, they make more of their own

called _____ . Fifth, living things get rid

of wastes.

Living things, also called _____ , are

made of cells. Cells are too _____ to see

with just your eyes. A tool called a microscope is used.

© Macmillan/McGraw-Hill

Classifying Living Things

Use your textbook to help you fill in the blanks.

How are living things classified?

1. Scientists place organisms into one of six groups,

   or _____ .

2. Organisms in the same kingdom have similar

   _____ .

How can organisms be grouped within a kingdom?

3. Traits are used to sort organisms into

   smaller _____ .

4. The smaller groups in a kingdom include:

   a. phylum

   b. _____

   c. _____

   d. _____

   e. _____

   f. species

What kinds of organisms have only one cell?

5. Three kingdoms that include organisms made up of one

   cell are _____ , _____ ,

   and _____ .

6. Tiny organisms are also called _____ .

7. Bacteria have no cell _____ and
   few cell parts.

8. A paramecium is a protist that has a structure which

   pumps extra _____ out of its cell.

9. Fungi have a cell nucleus and a _____,
just as plants do.

How are organisms named?

10. The scientific name for an _____ is made
up of a genus name and species name.

11. Scientists have named about 1.7 million _____

of organisms.

Critical Thinking

12. Using the chart on page 37 of your textbook, think of the
traits that an organism from a different order, family, genus,
and species might have. Name an animal from that species.

_____

_____

_____

_____

_____

_____

_____

_____

_____

© Macmillan/McGraw-Hill

Classifying Living Things

Match each word with its definition.

1. _____ protists

   a. a group of organisms with some members that make their own food (algae) and some that eat other organisms to live (paramecia)

2. _____ diseases

   b. organisms made up of cells with a cell wall and a nucleus but without chloroplasts

3. _____ trait

   c. the system used for identifying organisms

4. _____ fungi

   d. the group made up of one type of organism

5. _____ kingdom

   e. a characteristic of a living thing used to identify and classify it

6. _____ bacteria

   f. the largest group into which organisms can be classified

7. _____ species

   g. the smallest one-celled organisms

8. _____ classification

   h. the harmful effects of some microorganisms

© Macmillan/McGraw-Hill

Name _____ Date _____

Classifying Living Things

Fill in the blanks.

class	genus	similar
different	kingdom	six
family	one	species

Scientists study the traits of living things in order

to identify and classify them. Scientists divided Earth's

organisms into _____ groups. The largest

group, called a _____ , is divided into

smaller groups known as *phylum*, _____ ,

order, _____ , *genus*, and *species*.

Organisms in the same kingdom are somewhat

_____ to one another and are very

_____ from organisms in the other five

kingdoms. Each group in a kingdom becomes smaller and

smaller until only _____ type of orgranism

remains. The smallest group has only one type of organism

and is called a(n) _____ . Scientists use

_____ and _____ names to

identify individual types of organisms.

Red Tide: A Bad Bloom at the Beach

You're all ready for some fun and sun. But when you get to the beach, it's closed. Then you notice that the water is a strange color. You have to put your swimsuit away. Your beach is a victim of red tide!

Red tide isn't actually a tide. It is ocean water that is blooming with a harmful kind of algae. These one-celled organisms are poisonous to the sea creatures that eat them. The water isn't always red, either. Sometimes it's orange, brown, or green.

An outbreak of red tide can do a lot of damage. One outbreak on the coast of Florida killed tens of thousands of fish, crabs, birds, and other small animals within a few months. It also killed large animals like manatees, dolphins, and sea turtles. Red tides can make people sick if they eat infected shellfish.

Scientists are working to predict where and when red tides occur. They measure the amount of algae along coastlines. They use data collected from satellites to study wind speed and direction. This information helps scientists predict where blooms may develop. Scientists can then use their predictions to warn local agencies about incoming red tides.

Write About It

Infer What might you infer about a closed beach with reddish-colored water? How could the prediction of red tides be helpful to people?

What I Know

Complete each statement about red tides.

▶ A red tide is _____ that is blooming with

harmful _____ , _____ organisms.

▶ Red tides can make people sick if they _____
infected shellfish.

▶ A red tide in Florida killed tens of thousands of

_____ animals.

▶ Scientists are using _____ to collect data
about red tides.

What I Infer

Answer the questions by making inferences about red tides.

1. What might you infer about a closed beach with reddish-
   colored water?

   _____

   _____

2. How could the prediction of red tides be helpful to people?

   _____

   _____

   _____

The Plant Kingdom

Use your textbook to help you fill in the blanks.

How do we classify plants?

1. We can classify plants in _____ groups:

   those with and those _____ roots, stems, and leaves.

2. Plants without roots, stems, or leaves are

   _____ .

How do plants get what they need?

3. Plants make their own food by using _____ from sunlight.

4. Plants take in water and nutrients from the soil through

   their _____ .

Why are leaves important?

5. Plants use energy to change carbon dioxide and water into

   food, called _____ .

6. Plants get carbon dioxide and release water through openings on the undersides of their leaves, called

   _____ .

7. In a process called _____ , water exits the leaves of a plant.

8. Photosynthesis _____ food energy, and

   respiration _____ energy.

What are mosses and ferns?

9. Mosses and ferns are plants that use _____ to reproduce.

10. A spore case protects the spore from too much

_____ or too little _____ .

How do we use plants?

11. We use plants and plant parts such as bulbs, tubers,

_____ , _____ ,

_____ , and flowers for _____ .

12. We also use plants for _____ and

_____ .

Critical Thinking

13. If your family could grow only one kind of plant, which plant would be best for your family to grow?

_____

_____

_____

_____

_____

_____

_____

_____

The Plant Kingdom

Match the correct letter with the description.

a. epidermis	**d.** root	**g.** stem
b. photosynthesis	**e.** root hairs	**h.** stomata
c. respiration	**f.** spore	**i.** transpiration

1. _____ the part of a plant that carries food and nutrients to and from the roots and leaves

2. _____ a cell in a seedless plant that can grow into a new plant

3. _____ tiny holes found on the underside of a leaf

4. _____ a process that breaks down food and releases energy

5. _____ threadlike cells on a root that take in water and nutrients from the soil

6. _____ a plant part that takes up water and nutrients from the ground and holds the plant in place

7. _____ the thin, protective covering on a leaf that keeps water in the leaf

8. _____ the process that a plant uses to produce plant sugars from water, carbon dioxide, and energy from sunlight

9. _____ a process in which water exits a plant leaf

The Plant Kingdom

Fill in the blanks.

energy	organisms	processes
food	photosynthesis	respiration
leaves	plants	

Plants have the same needs as other living things.

Plants need air, water, _____ , and a

place to live. Plants get the energy they need in order to

grow from the _____ they make in their

_____ .

Photosynthesis and respiration are two very

important _____ that happen in

_____ . During _____ ,

energy is stored as plant sugars. During _____ ,

energy is released for use by the plant. All _____ ,

not just plants, depend on respiration for survival. All

animals also depend on the food made by plants.

How Seed Plants Reproduce

Use your textbook to help you fill in the blanks.

How do we classify seed plants?

1. Most plants that have _____ ,

   _____ , and _____
   produce seeds.

2. A seed is an undeveloped _____ inside a
   protective covering.

How do seeds form?

3. Pollination occurs when _____ travels
   from the male part of the flower to the female part of the
   flower.

4. When the male and female sex cells join together,

   fertilization takes place and a _____
   is formed.

How do seeds grow?

5. Seeds need _____ and the right
   conditions in order to start growing.

6. Seeds move from place to place carried by the

   _____ , attached to an animal's

   _____ , or passing through an animal's

   _____ and being left on the ground.

How are plants alike and different from their parents?

7. Inherited _____ are passed on from parent plants to offspring. Examples include the shapes of leaves and the colors of flowers.

8. Using inherited traits can help _____ grow bigger or better plants.

What are other ways plants can reproduce?

9. Plants can reproduce in ways other than through

_____ , _____ , or

_____ .

10. Examples of other ways that plants reproduce are

_____ , which are stems that grow

along the ground, and _____ and

_____ , which grow underground.

Critical Thinking

11. If an apple is dropped in a river, will its seeds ever grow?

_____

_____

_____

_____

_____

_____

How Seed Plants Reproduce

What am I?

Choose a word from the word box below that answers
each question.

a. fertilization	**c.** life cycle	**e.** pollination	**g.** seed
b. germination	**d.** ovary	**f.** reproduction	

1. I am the plant part that contains an undeveloped plant and
a food supply inside my protective coat.

   What am I? _____

2. I am the process that uses flowers and fruits to make new

   plants. What am I? _____

3. I am the part of a flowering plant in which the eggs are

   made and stored. What am I? _____

4. I am the process that occurs when pollen is moved from

   the anther to the pistil of a flower. What am I? _____

5. I am the process that occurs when the male and female sex

   cells of a flower join together. What am I? _____

6. I am the ongoing life story of a type of flowering plant.

   What am I? _____

7. I am the process during which a seed begins to grow.

   What am I? _____

Name _____ Date _____

How Seed Plants Reproduce

Fill in the blanks.

fertilization	ovary	pollination	seed
germinate	pollen	reproduction	

The plant kingdom has many members. One very large

group is made up of _____ plants. In this

type of plant, flowers, fruits, and cones are needed for

_____ . These plant parts contain male

and female plant parts that produce special cells. The male

cells are found in _____ on the anther

of a flower. The female cells, called eggs, are found in the

_____ at the base of the pistil.

Pollen can be transferred from the anther to the pistil

by the wind or by another organism, such as a honeybee,

through _____ . The male cells pass through

a tube into the ovary and join with the eggs, a process called

_____ . Fertilization in plants forms a seed.

The seeds _____ when conditions are right.

The seed produces a seedling with traits inherited from its

parent plant.

Dandelions and Me

Write About It

Personal Narrative Think about a time you were sad and someone said something that made you feel better. Write a personal narrative about the event. Tell how it made you feel.

Getting Ideas

Picture the event in your mind. Jot down what happened in the chart below. Start with what happened first.

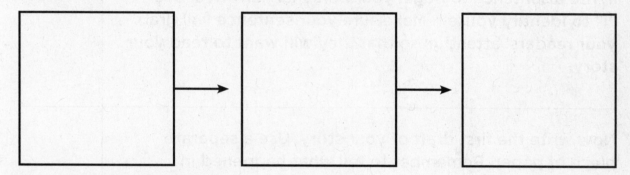

Planning and Organizing

Zoe wanted to write about the time she saw squirrels raiding the birdfeeder for sunflower seeds. Below are three sentences she wrote. Write "1" in front of the sentence that should be first. Write "2" in front of the sentence that should come next. Write "3" in front of the sentence that should be last.

_____ At first, some chickadees and blue jays came to the feeder and ate the seeds.

_____ After a few days, squirrels raided the feeder and carried seeds away.

_____ Last spring, we filled the birdfeeder in our backyard with sunflower seeds.

Revising and Proofreading

Here is a part of Zoe's personal narrative. Proofread it.
She made six errors. Find the errors and correct them.

I laughed when I saw that Rascal of a squirrel run off
with some seeds. I watched as the squirrel planted them. It
would fantastic if they grew? I waited about a weak. Every
day, I looked to see if the seeds had sprooted. Finally, there
was a little seedling, I was so excited.

Drafting

Write a sentence to begin your personal narrative. Use
"I" to identify youself. Make sure your sentence will grab
your readers' attention so that they will want to read your
story.

_____

Now write the first draft of your story. Use a separate
piece of paper. Remember to tell what happened in
sequence and to use time-order words.

Revising and Proofreading

Now revise and proofread your writing. Ask yourself:

▶ Did I use the pronoun "I" to tell my story?

▶ Did I tell what happened in sequence?

▶ Did I use time-order words?

▶ Did I correct all of the mistakes?

Kingdoms of Life

Choose the letter of the best answer.

1. A group of cells that do the same job forms a(n)

   a. organ system.

   b. organ.

   c. cell.

   d. tissue.

2. The protective covering on a leaf is its

   a. epidermis.

   b. stomata.

   c. root hairs.

   d. seed covering.

3. Into how many kingdoms do scientists place organisms?

   a. five

   b. six

   c. seven

   d. eight

4. The tiny holes on the underside of a leaf are the

   a. chlorophyll.

   b. epidermis.

   c. stomata.

   d. seeds.

5. How many basic jobs do living things perform?

   a. two

   b. three

   c. four

   d. five

6. A single cell that can grow into a new plant is called a

   a. spore.

   b. cone.

   c. stem.

   d. root.

7. What are the threadlike cells on a root?

   a. seeds

   b. root hairs

   c. runners

   d. cuttings

8. Tissues that form a group are called a(n)

   a. organ.

   b. cell.

   c. organ system.

   d. cell wall.

Name _____ Date _____

Choose the letter of the best answer.

9. In which process does water exit from a leaf?

   a. fertilization

   b. pollination

   c. germination

   d. transpiration

10. The smallest group in a kingdom is called a

   a. phylum.

   b. species.

   c. order.

   d. class.

11. The process by which plants release energy from food is called

   a. transpiration.

   b. respiration.

   c. germination.

   d. pollination.

12. Which of the following do scientists use to name organisms?

   a. genus and species

   b. phylum and class

   c. family and order

   d. order and genus

13. How is a plant cell different from an animal cell?

   a. Only plant cells contain cytoplasm.

   b. Only animal cells contain a nucleus.

   c. Only plant cells contain chloroplasts.

   d. Only animal cells contain mitochondria.

14. What do scientists use to see one-celled organisms?

   a. microscope

   b. test tube

   c. balance scale

   d. tongs

15. The joining of male cells and female cells in a plant is called

   a. pollination.

   b. germination.

   c. respiration.

   d. fertilization.

Name _____ Date _____

The Animal Kingdom

Complete the concept about the animal kingdom.

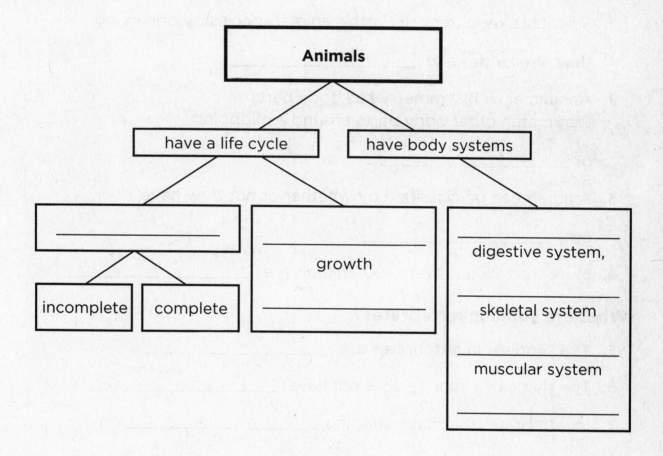

Animals Without Backbones

Use your textbook to help you fill in the blanks.

What are invertebrates?

1. Scientists keep track of Earth's animal species by observing

   their similarities and _____ .

2. An animal with symmetry has body parts
   that match other body parts around a midpoint

   or _____ .

3. Animals can be classified by whether or not they have

   a(n) _____ .

4. More than 95 out every 100 animals are _____ .

What are some invertebrates?

5. The simplest invertebrates are _____ .

6. The shape of a sponge does not have _____ .

7. Invertebrates that have stinging _____ on

   their tentacles are called _____ .

8. Clams, squid, and snails are soft-bodied invertebrates with

   hard shells and are called _____ .

9. Sea stars, sea urchins, and sand dollars are spiny-skinned

   invertebrates, called _____ .

10. All echinoderms have a support structure inside their

    bodies, called a(n) _____ .

© Macmillan/McGraw-Hill

What are arthropods?

11. Invertebrates with jointed legs and body sections are

called _____ .

12. Arthropods have a hard outer covering, called a(n)

_____ , that protects their bodies and

holds in moisture.

How are worms classified?

13. Worms are classified as flatworms, _____ ,

or _____ worms.

14. Flatworms have ribbon-like bodies, and some types

_____ inside the bodies of other animals.

15. Roundworms have thin bodies with _____
ends.

Critical Thinking

16. Why do you think the first way an animal is classified is by
whether it is a vertebrate or an invertebrate?

_____

_____

_____

_____

_____

_____

Animals Without Backbones

Choose the word from the box below that completes
each sentence.

arthropod	echinoderm	exoskeleton	mollusk
cnidarian	endoskeleton	invertebrate	sponge

1. The hard outer covering that protects an invertebrate's

   body is its _____ .

2. A spiny-skinned invertebrate, such as a sea star, is

   called a(n) _____ .

3. A(n) _____ is an animal without
   a backbone.

4. A soft-bodied invertebrate, such as a clam or snail, is

   called a(n) _____ .

5. An invertebrate with jointed legs and a sectioned body

   is a(n) _____ .

6. An internal support structure in an animal is

   a(n) _____ .

7. An invertebrate with poisonous stingers is

   a(n) _____ .

8. The simplest kind of invertebrate is a(n) _____ .

Animals Without Backbones

Fill in the blanks.

arthropods	endoskeleton	mollusks
backbone	exoskeleton	segmented
cnidarians	invertebrates	sponges

Scientists use various traits to classify Earth's two million animals. One way to classify animals is as vertebrates or _____ . Vertebrates have a(n) _____ , and invertebrates do not.

Insects and arachnids are invertebrates, called _____ , that have a hard outer _____ . Echinoderms are invertebrates that have a(n) _____ inside their bodies. The simplest invertebrates are _____ . Other invertebrate groups are _____ with poisonous stingers and _____ with soft bodies protected by hard shells. The final group of invertebrates is worms: flatworms, roundworms, and _____ worms. Some kinds of flatworms and roundworms live inside the bodies of other animals.

Animals With Backbones

Use your textbook to help you fill in the blanks.

What are vertebrates?

1. Vertebrates are animals that have a(n) _____ .

2. Some vertebrates are _____ and
   maintain their body temperature by breaking down food to

   get _____ .

3. Some vertebrates, such as fish, amphibians, and reptiles,

   are _____ and cannot keep constant

   body _____ .

4. There are seven classes of vertebrates: mammals,

   _____ , reptiles, birds, and three classes

   of _____ .

5. The three classes of fish are jawless fish, cartilaginous fish,

   and _____ fish.

What are some other vertebrate groups?

6. A vertebrate that spends part of its life in water and part

   on land is called a(n) _____ .

7. Snakes, lizards, turtles, and crocodiles are _____ .

   They have tough, dry, scaly _____ that holds
   in moisture.

8. The only animals that have feathers are _____ .

   Like mammals, they are also _____ .

© Macmillan/McGraw-Hill

What are mammals?

9. A warm-blooded vertebrate with hair or fur is

   a(n) _____ .

10. Mammals can live on land, in trees, and in

    _____ .

11. Most mammals give birth to live young. Only a

    few _____ .

12. Kangaroos, koalas, and opossums carry their young in

    _____ until they are grown.

13. The platypus and spiny anteater are the only mammals that

    reproduce by _____ .

Critical Thinking

14. Why do you think there are three separate groups of fish
    instead of one group for all fish?

    _____

    _____

    _____

    _____

    _____

    _____

Name _____ Date _____

Animals With Backbones

What am I?

Choose a word from the word box below that answers each question.

amphibian	cartilage	reptile	warm-blooded
bird	cold-blooded	vertebrate	

1. _____ Animals in my group have backbones. What am I?

2. _____ I can keep my body at a constant temperature. I do this by breaking down food to release heat. What am I?

3. _____ My body temperature changes with the surrounding temperature. What am I?

4. _____ I spend part of my life in water and part on land. My skin must be kept moist. What am I?

5. _____ Snakes and lizards are part of my group. We live on land and have tough, scaly skin. What am I?

6. _____ I have scales and feathers and hollow bones that make my body light enough to fly. What am I?

7. _____ I am rubbery and make up skeletons in lampreys, sharks, and rays. What am I?

Animals With Backbones

Fill in the blanks.

bony	invertebrates	surroundings
cold-blooded	mammals	warm-blooded
fur	reptiles	vertebrates

Invertebrates are the largest group of animals on Earth.

The second-largest group is _____ .

Vertebrates have backbones, and _____

do not.

Some vertebrates keep their bodies at one temperature

by eating food to release heat. These animals are

_____ and are birds or _____ .

Birds have feathers, and mammals have _____

or hair. Other vertebrates are _____

and their body temperature depends on their

_____ . Cold-blooded animals are

classified as jawless fish, cartilaginous fish, _____

fish, amphibians, or _____ .

Name _____ Date _____

Gentle Giants

Write About It

Explanatory Writing Find out more about another endangered animal. Write a short explanation of why it is endangered.

Getting Ideas

Select an endangered animal. Use the cause-and-effect chart below. Fill it in as you do research.

Cause	→	Effect
	→	
	→	
	→	
	→	

Planning and Organizing

Here are some sentences that Kristen wrote about lemurs. Circle the part of the sentence that tells the cause. Underline the part that shows the effect.

1. Lemurs are losing their habitat because people cut down trees for farming.

2. Some people hunt lemurs because they are afraid of them.

3. Every year there are fewer lemurs because they are hunted for food.

© Macmillan/McGraw-Hill

Revising and Proofreading

Here is part of Kristen's explanation. She made six capitalization mistakes. Find the mistakes and correct them.

> Lemurs live on madagascar and the comoro islands. These are islands off the coast of africa. Before humans arrived, there were many species of Lemurs. over time, at least fourteen species became extinct.

Drafting

Write a sentence to begin your explanation. Tell the name of the animal and your main idea about it.

_____

_____

Now write your explanation. Use a separate piece of paper. Begin with your topic sentence. Include facts and details to explain how the animal became endangered. End by telling what scientists are doing to save this animal.

Revising and Proofreading

Now revise and proofread your writing. Ask yourself:

▶ Did I explain how the animal became endangered?

▶ Did I tell what scientists are doing to save it?

▶ Did I correct all mistakes?

Name _____ Date _____

Systems in Animals

Use your textbook to help you fill in the blanks.

How do animals move and sense changes?

1. The skeletal system is made of bones that

   _____ soft body organs and work with the

   body's _____ system.

2. The muscular system is made of muscles that

   move _____ .

3. Earthworms move by _____ and
   stretching their muscles.

4. Sight, hearing, taste, touch, and smell are senses that help

   an animal detect _____ in its surroundings.

5. An animal's body systems are controlled by

   its _____ .

How do air and blood travel in the body?

6. An animal gets the _____ it needs to live

   through the _____ system.

7. This system brings oxygen to the blood and

   removes _____ .

8. Animals take in oxygen in several ways: through their

   tissues, _____ , or _____ .

9. The heart, blood, and blood vessels are organs that make

   up an animal's _____ system.

© Macmillan/McGraw-Hill

10. Blood is pumped by the _____ to all of the cells in an animal's body.

11. The bladder and kidneys are organs in the _____ system that remove _____ from the body.

How is food broken down?

12. Animals use food for _____ .

13. After food has been eaten, it must be _____ to release its nutrients.

14. The _____ system breaks down food.

15. The _____ churns and mixes food.

Critical Thinking

16. Do you think that an organism's nervous system is affected by what it eats?

_____

_____

_____

_____

_____

_____

_____

Name _____ Date _____

Systems in Animals

Fill in the blank. Then circle the term in the puzzle.

1. These two systems work together to move the body.

_____

_____

2. The brain, spinal cord, nerves, and sense organs are part of this system.

_____

3. Oxygen and waste gases are exchanged through this system.

_____

4. The heart, blood, and blood vessels make up this system.

_____

5. Any organ that removes wastes from the body is part of this system.

_____

6. Food is broken down to release nutrients in this system.

_____

```
C I R C U L A T O R Y
T B S U O V R E N E I
M S M T E M C V V X R
K K U B V L M I E C V
R E S P I R A T O R Y
B L C A T B T S T E P
N E U M S R O E I T E
F T L C E I K G R O G
L A A K G V G I C R K
C L R F I N M D V Y Y
```

Name _____ Date _____

LESSON
Cloze Activity

Systems in Animals

Fill in the blanks.

circulatory system	nervous systems	skeletal system
excretory system	oxygen	wastes
muscular system	sense	

An animal's organ systems help it to meet its needs

and respond to changes. The _____

controls an animal's organ systems and contains the

_____ organs that detect changes. The

bones of the _____ protect the soft organs.

It also works with the _____ to help an

animal move. The digestive system breaks down food into

nutrients.

Animals need _____ , which comes

through the respiratory system. The heart, blood, and

blood vessels make up the _____ . The

blood carries food and oxygen to the cells, and it carries

_____ to the _____ . In this

system, wastes are removed by the kidneys and lungs.

© Macmillan/McGraw-Hill

Chapter 2 • The Animal Kingdom
Reading and Writing

Use with **Lesson 3**
Systems in Animals   **39**

Name _____ Date _____

Animal Life Cycles

Use your textbook to help you fill in the blanks.

What are the stages of an animal's life?

1. Organisms go through changes in their lives, including

   _____ , growth, _____ ,

   and death.

2. A penguin chick depends on its parents for warmth,

   shelter, and _____ .

3. An organism's _____ is how long it can
   usually live in the wild.

What is metamorphosis?

4. The process of _____ includes a series
   of separate growth stages.

5. _____ includes separate growth
   stages that have changes that are hard to see.

   _____ includes growth stages that are
   different at every stage.

How do animals reproduce?

6. All life cycles include _____ .

7. Parent animals make _____ .

8. Budding and _____ are types of reproduction with one parent.

9. Organisms that reproduce with one parent produce exact

copies, or _____ .

10. In two-parent reproduction, a male sperm cell and a female

egg cell combine during _____ and

produce a(n) _____ .

11. Traits such as eye color, height, and body color are

determined by _____ before an organism
is born.

What is inherited?

12. An _____ is a set of actions that parents

pass on to their _____ .

Critical Thinking

13. Why do you think that different animals have different
life cycles?

_____

_____

_____

_____

_____

_____

Name _____ Date _____

Animal Life Cycles

What am I?

Choose the letter that matches the word from the word box below to answer each question.

a. clone	**d.** learned behavior	**g.** metamorphosis
b. heredity	**e.** life cycle	
c. instinct	**f.** life span	

1. I am the stages through which an organism passes, including birth and reproduction. What am I? _____

2. I am the length of time that an organism is expected to live. What am I? _____

3. I am the process that takes place in a series of separate growth stages. What am I? _____

4. I am the offspring of only one parent. I am an exact copy of my parent. What am I? _____

5. I control the traits that are passed on from parent to offspring. What am I? _____

6. I am the behavior with which an organism is born. What am I? _____

7. I am a behavior that an organism gains from experience. What am I? _____

© Macmillan/McGraw-Hill

Animal Life Cycles

Fill in the blanks.

| birth | growth | metamorphosis | separate |
| gradual | life span | produce | |

All animals go through stages that make up the life

cycle. These stages include _____ ,

_____ , reproduction, and death. The

amount of time an animal is expected to live is called its

_____ . An animal is expected to live long

enough to _____ offspring.

The stages of growth can be _____

or _____ and different, a process called

_____ . The life cycle of every animal begins

with birth and ends with death.

Name _____ Date _____

Meet Christopher Raxworthy

Read the passage in your textbook. Look for information about the Mantella poison frog and dwarf dead leaf chameleon.

 Write About It

Compare and Contrast How does the life cycle of the Mantella poison frog compare to the life cycle of the dwarf dead leaf chameleon?

Compare and Contrast

Fill in the Compare and Contrast graphic organizer. Tell how the frog and chameleon are alike and how they are different. Then, answer the question.

Frog	Chameleon	Frog and Chameleon
Its body has vivid colors to warn _____.	Its body resembles a(n) _____.	Babies hatch from _____.
Females lay eggs in _____ areas.	The animal hides during the day in dead leaves on the _____.	Frogs and Chameleons become _____ in about _____.
Eggs hatch when it _____.	Females lay eggs in _____.	
Tadpoles move to a nearby _____.	Eggs hatch in _____ weeks.	

Compare and Contrast

Read the paragraph below. Compare and contrast the work of Christopher Raxworthy and the scientists in Madagascar with that of the scientists at the San Diego National Wildlife Refuge.

San Diego National Wildlife Refuge

In the 1990s, the people of San Diego began working with government groups to help protect the environment. A wildlife refuge was created. The goals of the San Diego refuge include preserving endangered species and helping endangered species increase in number. The refuge protects all the wildlife native to the area, not just the endangered species. It also protects the habitats of migratory birds. The refuge provides visitors with opportunities to learn about wildlife.

 Write About It

Write a short paragraph in which you compare and contrast the goals of Christopher Raxworthy and the other scientists in Madagascar with those of the scientists at the San Diego Refuge Complex.

_____

_____

_____

_____

_____

The Animal Kingdom

Circle the letter of the best answer.

1. An animal that lives part of its life in water and part of it on land is a(n)

    a. amphibian.

    b. reptile.

    c. mammal.

    d. fish.

2. What is an arthropod?

    a. an invertebrate with a spiny skin that lives in the ocean

    b. an invertebrate that remains anchored to one spot

    c. an invertebrate that lives inside the body of another animal

    d. an invertebrate with jointed legs and a body divided into sections

3. The only warm-blooded animals with a body covering of feathers are

    a. snakes.

    b. birds.

    c. mammals.

    d. fish.

4. The circulatory system is made up of the

    a. brain, spinal cord, nerves, and sensory organs.

    b. heart, blood vessels, and blood.

    c. kidneys and lungs.

    d. mouth, stomach, and digestive juices.

5. An organism that is produced by only one parent organism is called a(n)

    a. egg.

    b. embryo.

    c. clone.

    d. seed.

6. The passing of traits from parents to their offspring is known as

    a. cloning.

    b. heredity.

    c. instinct.

    d. behavior.

Circle the letter of the best answer.

7. A cold-blooded animal
   a. cannot keep a constant body temperature.
   b. can keep a constant body temperature.
   c. uses the food it eats to make heat.
   d. has a short life span.

8. What is an instinct?
   a. a learned behavior
   b. a learned trait
   c. an inherited behavior
   d. an environmental trait

9. Which body system controls all of the other systems in an animal's body?
   a. nervous system
   b. digestive system
   c. skeletal system
   d. respiratory system

10. A hard protective outer covering that keeps in moisture is a(n)
    a. endoskeleton.
    b. exoskeleton.
    c. backbone.
    d. egg.

11. If an animal's body parts match around a midpoint or central line, that animal has
    a. endoskeleton.
    b. symmetry.
    c. exoskeleton.
    d. instinct.

12. Animals are classified as vertebrates if they
    a. do not have a backbone.
    b. have a backbone.
    c. can move.
    d. live on land.

Sea Otters: Key to the Kelp Forest
From *Ranger Rick*

Read the Unit Literature feature in your textbook.

Write About It

Response to Literature Research another place where plants and animals depend on each other. Write a report describing how the plants and animals interact.

_____

_____

_____

_____

_____

_____

_____

_____

_____

_____

_____

_____

_____

_____

_____

Exploring Ecosystems

Ecosystems

| contain living factors known as _____ | can be broken down into six different _____ | contain different types of organisms | contain nonliving factors known as _____ |

Examples of

_____

factors are:

1. _____

2. _____

3. microorganisms

The three main classifications of organisms are:

1. producers

2. _____

3. _____

The six

_____ are:

1. _____

2. _____

3. tropical rain forest

4. _____

5. _____

6. desert

Examples of

_____

factors are:

1. _____

2. _____

3. _____

4. _____

5. _____

Introduction to Ecosystems

Use your textbook to help you fill in the blanks.

What is an ecosystem?

1. The living and nonliving things in the _____

   make up a(n) _____ .

2. Plants, animals, and bacteria are living things, or

   _____ , in an environment.

3. Water, rocks, and soil are some of the nonliving things,

   or _____ , in an environment.

4. Ecosystems can be large or _____ .

5. An important abiotic factor in an ecosystem is

   its _____ , the typical weather pattern
   in an environment.

6. Living things in an ecosystem _____
   on nonliving things to survive.

7. The place in an ecosystem in which each organism lives

   is that organism's _____ .

8. Different ecosystems have _____
   types of habitats.

9. To lay its eggs, a frog depends on the

   _____ in a pond.

© Macmillan/McGraw-Hill

What are populations and communities?

10. Ecosystems have different _____ of species.

11. All of the populations in an ecosystem make up

a(n) _____ .

12. When scientists want to know about a(n) _____ , they look at its populations and communities.

13. Warm and moist ecosystems usually have larger

communities than _____ and

_____ ecosystems.

Critical Thinking

14. What do you think is the most important factor affecting the size of a community in an ecosystem?

_____

_____

_____

_____

_____

_____

_____

© Macmillan/McGraw-Hill

Introduction to Ecosystems

Read each definition. Write the term in the blank and fill in
the crossword puzzle.

Across

1. all the members of a
species in an ecosystem

_____

4. all of the populations
in an ecosystem

_____

5. a living thing's place
to live in an ecosystem

_____

7. an important abiotic
factor in all ecosystems

_____

Down

2. the nonliving factors of an
ecosystem, such as rocks

_____

3. all the living and nonliving
things in an environment

_____

6. the living factors of an
ecosystem, such as plants

_____

© Macmillan/McGraw-Hill

Introduction to Ecosystems

Fill in the blanks.

abiotic factors	climates	habitats
bacteria	dry	small
biotic factors	ecosystem	

All of the living and nonliving things in an area make up the environment. An environment's living things, such as plants, animals, and _____ , are called _____ . Nonliving things, such as water, rocks, and soil, are called _____ . The biotic and abiotic factors in an environment work together to form a(n) _____ .

Ecosystems can be large or _____ . They can also have very different _____ . Some ecosystems are hot and _____ , and others are cold and wet. Different ecosystems have _____ that are suited to different types of living things. For example, a desert community is suited to cacti and lizards.

Biomes

Use your textbook to help you fill in the blanks.

What is a biome?

1. A large ecosystem with a unique set of characteristics is

   called a(n) _____ .

2. Biomes have distinct patterns of _____

   and _____ . Some biomes stretch across

   an entire _____ .

What are grasslands and forests?

3. A biome whose plant life includes mostly grasses growing

   in its _____ soil is a(n) _____ .

4. During hot, dry summers, _____ burn,
   and this produces rich soil for farming.

5. Oaks, maples, and hickories in the _____
   forests lose their leaves each year.

6. The _____ is a hot and humid biome with
   plenty of rainfall.

What are deserts, taiga, and tundra?

7. Black bears live in the _____ , the largest
   biome in the world.

8. Cacti and yucca survive in the _____ where

   there is very little _____ .

9. The tundra is home to mammals that _____ through the winter and plants that grow close to the

frozen _____ .

Are there water biomes?

10. Water _____ are grouped differently from land biomes.

11. Freshwater ecosystems include lakes, ponds, rivers,

_____ , and some _____ .

12. Saltwater ecosystems include _____ , and

_____ where freshwater and saltwater ecosystems meet.

Critical Thinking

13. Why do you think so many different organisms live in the tropical rain forest?

_____

_____

_____

_____

_____

_____

_____

_____

Biomes

Match the correct word with its description.

a. biome	**d.** grassland	**g.** tundra
b. deciduous forest	**e.** taiga	
c. desert	**f.** tropical rain forest	

1. _____ This is a large ecosystem that has its own special plants, animals, soil, and climate.

2. _____ This is a biome, such as a prairie, that has fertile soil and some, but not much, rainfall.

3. _____ Many trees in this biome lose their leaves every year in autumn.

4. _____ This biome is located near the equator. It is hot and humid year round. It is home to a large variety of plants and animals.

5. _____ Earth's northern regions are the location of this forest biome. Its plant life includes conifers, lichens, and mosses.

6. _____ This dry biome gets little rain.

7. _____ The ground is frozen year round in this cold, dry biome.

Biomes

Fill in the blanks.

biomes	dry	taiga
characteristics	harsh	tropical rain forest
cold	organisms	
deciduous forest	populations	

Earth has six major ecosystems. These ecosystems,

also known as _____ , have their own

_____ , including temperature,

precipitation, and soil. Each biome also has a special

community made up of different _____

of plants and animals.

The _____ has trees that lose their leaves

each autumn. The _____ has a climate

suitable to the greatest variety of _____ .

Other biomes have _____ climates.

The _____ and tundra have very

_____ environments. Deserts have

very _____ environments. There are

fewer plants and animals in these three biomes.

Name _____ Date _____

Museum Mail Call

Read the selection from your textbook. Look for
information about how building affects an ecosystem.
On a separate piece of paper, write the sentences that
state facts about the mangrove swamp.

Write About It

Draw Conclusions What might happen to the plants
and animals of Florida's wetlands if people continue to
build there?

Fill in the Draw Conclusions graphic organizer about
the mangrove swamp.

My Prediction	What Happens
Many mangroves are being replaced by stores, _____ , and parking lots.	Cutting down the mangrove trees will change the _____ .
The mangroves are home to many _____ .	Loss of the mangroves will affect the population of the _____ .
Mangrove roots provide shelter for _____ .	Animals will have to find a new habitat, and some species may not _____ .
The mangroves protect the _____ from wind, waves, and floods.	The coast will not be protected from winds, waves, and _____ .

© Macmillan/McGraw-Hill

Reread Tommy's message. If you were one of the museum's scientists, how would Tommy's note help you? What would you and other museum scientists do to keep the mangroves safe? How would you protect the plants and animals that live in the mangroves? Write an informative response to Tommy and answer his question.

TO: Tommy

FROM: American Museum of Natural History

SUBJECT: Save the Mangroves!

Dear Tommy,

_____

_____

_____

_____

_____

_____

_____

_____

_____

_____

_____

Relationships in Ecosystems

Use your textbook to help you fill in the blanks.

How do organisms depend on one another?

1. Organisms in an ecosystem depend on producers

   for _____ .

2. Plants are producers that make food using energy

   from _____ .

3. Energy moves from producers to _____ ,
   organisms that need to eat other organisms for energy.

4. The three types of consumers are herbivores, which eat

   only _____ , carnivores which eat only

   other _____ , and _____ ,
   which eat both plants and animals.

5. Ecosystems also have organisms, called _____ ,
   that break down and recycle plant and animal remains.

What is a food chain?

6. The order in which energy passes through organisms in

   an ecosystem is called a(n) _____ .

7. Algae and green plants are first in a _____
   food chain.

8. Bacteria and other _____ break down
   dead organisms.

What is a food web?

9. Food chains in an ecosystem are connected in

   a(n) _____ .

10. Food webs show the relationships between

    _____ , which are organisms that hunt for

    food, and _____ , which are organisms
    that are hunted.

11. Plants in a food web compete for _____

    and _____ .

What is an energy pyramid?

12. Energy in an ecosystem travels from the producers to the

    _____ and then to the omnivores and
    carnivores.

13. A model of how much energy there is in an ecosystem

    is called a(n) _____ .

Critical Thinking

14. Where do you think decomposers fit into the energy
    pyramid?

    _____

    _____

    _____

    _____

    _____

Relationships in Ecosystems

What am I?

Choose a word from the word box below that answers
each question.

a. competition	**d.** energy pyramid	**g.** producer
b. consumer	**e.** food chain	
c. decomposer	**f.** food web	

1. _____ I am a living thing that can use energy from
the Sun to make food. What am I?

2. _____ I am a living thing that must use other
organisms as food to get energy. What am I?

3. _____ I am an organism that breaks down and
recycles the remains of dead organisms in an
ecosystem. What am I?

4. _____ I show the order, or sequence, in which
organisms in an ecosystem consume one
another. What am I?

5. _____ I am the struggle that takes place among
organisms for food, water, and other things
needed to live. What am I?

6. _____ I am formed when food chains are linked
together. What am I?

7. _____ I show how much energy there is at each step
of a food chain in an ecosystem. What am I?

© Macmillan/McGraw-Hill

Relationships in Ecosystems

Fill in the blanks.

carnivores	energy pyramid	groups	producers
competition	food chain	live	
decomposers	food web	organisms	

An ecosystem is a community of organisms that are in

_____ with each other for limited

amounts of water, food, energy, and space.

Members of an ecosystem can be sorted into three

main groups: _____ , consumers, and

_____ . The order, or sequence, in

which _____ eat one another is called

a(n) _____ . Different food chains can

be connected to form a(n) _____ .

Energy moves through an ecosystem from plants to

herbivores and then to _____ . An

energy pyramid shows how energy is used in an

ecosystem. Without producers, consumers could not

_____ .

© Macmillan/McGraw-Hill

Name _____ Date _____

The Moth that Needed the Tree

Write About It

Expository Writing Research another example of how insects and plants depend on each other. Write a report with facts and details from your research.

Getting Ideas

Think about what you learned in this chapter and through your research. Fill in the chart below. Tell the main idea and two details about yucca trees and yucca moths.

Main Idea	Details

Planning and Organizing

Mireya wrote three sentences. Write "Main Idea" next to the sentence that tells the main idea. Write "Detail" next to each sentence that tells a detail.

1. _____ In spring, yucca moths crawl out of their cocoons.

2. _____ The yucca moth and yucca tree need each other to live.

3. _____ Female moths gather pollen from the yucca tree.

Revising and Proofreading

Mireya wrote some sentences. She did not include many details. Choose a word or set of numerals from the box. Write it on the line.

3-5	black	few	8-10

A yucca moth is about _____ mm

long. Its color is _____ . In only a

_____ days, the female moth places about

_____ eggs in a yucca flower's ovary.

Drafting

Begin your report. Start with a topic sentence. Tell the main idea of your report.

_____

Now write your report. Use a separate piece of paper. Start with the sentence you wrote above. Include facts and details about the yucca tree and the yucca moth. At the end of your report, draw a conclusion about how they help each other.

Revising and Proofreading

Now revise and proofread your writing. Ask yourself:

▶ Did I include facts and details?

▶ Did I draw a conclusion at the end of the report?

▶ Did I correct all mistakes?

© Macmillan/McGraw-Hill

Name _____ Date _____

Exploring Ecosystems

Circle the letter of the best answer.

1. Water, rocks, and other nonliving things in an environment are called

   a. biotic factors.

   b. abiotic factors.

   c. a population.

   d. an ecosystem.

2. What do all of the living and nonliving things in an environment make up?

   a. a species

   b. a population

   c. an ecosystem

   d. a community

3. Each plant and animal in an ecosystem has its own place to live. That is the organism's

   a. habitat.

   b. location.

   c. biome.

   d. abiotic factor.

4. All of the barrel cacti in a desert make up a group of organisms called a(n)

   a. ecosystem.

   b. population.

   c. community.

   d. habitat.

5. All of the cacti, insects, birds, and lizards in the desert are part of the desert

   a. habitat.

   b. population.

   c. community.

   d. producers.

6. The living things in an environment are called

   a. biotic factors.

   b. abiotic factors.

   c. a population.

   d. an ecosystem.

7. What is the name for the six major ecosystems on Earth?

   a. biomes

   b. biotic factors

   c. abiotic factors

   d. ecology

© Macmillan/McGraw-Hill

Circle the letter of the best answer.

8. Which biome has different types of trees, most of which lose their leaves in the autumn?

   a. desert

   b. tundra

   c. tropical rain forest

   d. deciduous forest

9. Which hot, humid biome is located near Earth's equator?

   a. desert

   b. tundra

   c. tropical rain forest

   d. deciduous forest

10. Which of the following biomes has the driest climate?

    a. desert

    b. deciduous forest

    c. grassland

    d. taiga

11. Which of these makes food in an ecosystem?

    a. producers

    b. consumers

    c. decomposers

    d. herbivores

12. Which statement best describes a consumer?

    a. Consumers make their own food.

    b. Consumers cannot make their own food.

    c. Consumers get energy from the Sun.

    d. Consumers recycle the remains of dead organisms.

13. The struggle among organisms for food, water, and other needs is called

    a. competition.

    b. a food web.

    c. a food chain.

    d. a predator.

14. What relationship is shown below?

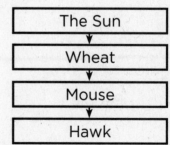

    a. food web

    b. food chain

    c. energy pyramid

    d. food pyramid

Name _____ Date _____

Surviving in Ecosystems

Use your textbook to help you fill in the blanks.

Changes in Ecosystems

Cause	⟶	Effect
Natural Events ⟶		▶ short-term changes, like the change of _____ ▶ long-term changes that result from a(n) _____ or the eruption of a(n) _____
Living Things ⟶		▶ harmful changes, like _____ ▶ helpful changes, like _____
People ⟶		▶ helpful changes, like _____ ▶ harmful changes, like _____ , _____ , and endangering living things

Animal Adaptations

Use your textbook to help you fill in the blanks.

What are adaptations?

1. Survival is not easy for organisms, because each

   ecosystem has special _____ .

2. Organisms have traits or _____ that
   help them survive in their environments.

3. A bird's beak and a camel's hump are two examples

   of adaptations that help animals to _____ .

4. Organisms that live in hot desert ecosystems have

   adaptations for staying _____ and

   saving _____ .

5. The fennec fox has large ears that give off _____

   and thin _____ that helps it stay cool.

6. Kangaroo rats survive in the desert because they get

   water from the _____ they eat.

7. Camels have humps to store fat for _____ ,

   and they have _____ to walk on sand.

What are some other adaptations of animals?

8. Animals living in different _____ have
   different adaptations.

9. Some animals have adaptations, like the _____ on a hedgehog, to protect themselves

from _____ .

10. Some animals survive because they blend in with the colors and shapes in their environments, an adaptation called _____ .

11. Hover flies and scarlet king snakes can survive because they look like other, more dangerous organisms, an

adaptation called _____ .

12. Animals can avoid cold winters by _____ or resting until the weather gets warmer.

How else do animals survive?

13. Living things _____ in harmful and

_____ ways.

Critical Thinking

14. What kinds of adaptations do the Egyptian plover and the Nile crocodile have?

_____

_____

_____

_____

_____

_____

_____

Animal Adaptations

Choose a word from the word box below that correctly fills in the blank.

adaptation	hibernate	mimicry	traits
camouflage	migrate	survive	

1. Animals have adaptations that help them

   _____ in their ecosystem.

2. Some organisms "copy" the traits of other living things
   in their environment. This adaptation is

   called _____ .

3. Organisms have _____ that help them
   survive in their environments.

4. Any trait that helps an organism survive in its

   environment is called a(n) _____ .

5. Some animals periodically _____ to
   different locations for warmer weather.

6. The fur of an arctic fox changes color so it can
   blend into its environment. This adaptation is called

   _____ .

7. Some animals survive the cold winter because they are
   able to remain completely still for a long period of time,

   or _____ .

Name _____ Date _____

Animal Adaptations

Fill in the blanks.

adaptations	behavior	challenges	different	predators
avoid	camouflage	colors	mimicry	survive

All ecosystems present challenges to the

organisms that live there. Living things have different

_____ that make them better suited

to the _____ in their environments

and help them _____ .

Survival in _____ environments

requires different adaptations. An organism with

_____ can hide from _____

because it blends in with the _____

and shapes of its environment. An organism that has

_____ is copying the physical traits

and _____ of other organisms that

predators usually _____ . Different

animals have different adaptations and different

behaviors, but all of them have the same goal—survival.

© Macmillan/McGraw-Hill

Plants and Their Surroundings

Use your textbook to help you fill in the blanks.

How do plants respond to their environments?

1. Plants respond to changes in their _____ in many different ways.

2. Something in the environment that causes

   a living thing to react is called a(n) _____ .

3. The response of a plant to a stimulus is

   called a(n) _____ .

4. A plant responds to a stimulus by changing its

   _____ or _____ of growth.

5. Plant stems that grow upward, _____

   a source of light, and plant _____ that grow toward a source of water are tropisms.

6. Plant roots also grow downward, in the direction of the

   pull of _____ .

7. The green _____ of plants grow

   _____ , away from gravity.

What are some plant adaptations?

8. Like animals, plants have _____ for various environments.

9. A cactus in the desert has adaptations for conserving

   _____ , such as spongy tissue inside

   and a _____ , waxy outer cover.

10. Some trees lose their _____
    every winter.

11. The trees live on _____ food

    until spring, when new _____ grow
    and the tree begins making food again.

Critical Thinking

12. What do you think would happen to trees if their leaves
    did not fall off before winter?

    _____

    _____

    _____

    _____

    _____

    _____

    _____

    _____

    _____

Plants and Their Surroundings

Match the correct word with the description.

adaptation	light	tropism	water
energy	stimulus	upward	

1. A tree that loses its leaves in the fall survives
   during the winter by living on stored food

   for _____ .

2. A cactus has spongy tissue inside for storage and a
   very thick, waxy skin on the outside to prevent loss

   of _____ .

3. A trait that helps a plant survive in its environment is

   called a(n) _____ .

4. Anything in the environment that causes a plant to
   react, such as chemicals, heat, gravity, or water, is

   called a(n) _____ .

5. The reaction of plants to any stimulus is

   called _____ .

6. Some stimuli that affect plants are chemicals, heat,

   gravity, water, and _____ .

7. A plant responds to gravity in two ways: its roots grow
   downward, and its green stems

   grow _____ .

Name _____ Date _____

Plants and Their Surroundings

Fill in the blanks.

adaptations	leaves	stimulus
direction	light	tropisms
ecosystem	photosynthesizing	water
food	respond	

Plants, like animals, have traits that help them to

survive in their environments. Plants in a desert

_____ have _____ for

storing _____ . Deciduous trees lose

their _____ in the fall. They live on

stored _____ until the leaves grow

back in the spring and start _____ .

Plants cannot move, but they can _____

to stimuli. All plant responses are called _____ .

A plant can react to a(n) _____ by

changing its _____ or pattern of

growth. Plant roots respond to water, and plant stems

respond to _____ sources. Plant roots also

respond to the pull of gravity.

© Macmillan/McGraw-Hill

A Field of Sun

Write About It

Descriptive Writing Do some research about another plant. Write a description of how this plant reacts to its environment.

Getting Ideas

First, choose a plant. Write its name in the center circle in the web below. Do some research. Write details you find about this plant in the outer circles.

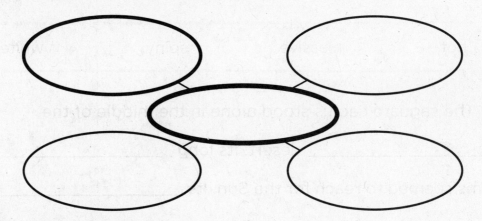

Planning and Organizing

Alberto decided to describe the saguaro cactus. Here are some sentences he wrote. Identify the sense to which the details in the sentence appeal. The five senses are sight, hearing, taste, smell, and touch.

1. _____ The saguaro cactus has a smooth, waxy skin.

2. _____ It has pretty white flowers with yellow centers.

3. _____ The cactus makes a sweet nectar.

Drafting

Write a sentence to begin your description. Tell what plant you are writing about.

_____

Now write your description. Use a separate piece of paper. Begin with the sentence you wrote above. Use vivid details and sensory words to describe the plant.

Revising and Proofreading

Here is part of Alberto's description. He left out some sensory words. Choose words from the box or pick your own. Write them in the blanks.

hot	massive	spiny	white

The saguaro cactus stood alone in the middle of the

_____ desert. Its long _____

arms seemed to reach for the Sun. Its _____

stem was about 20 inches in diameter. Its beautiful

_____ flowers waited for the Sun to

go down. Then they bloomed.

Now revise and proofread your writing. Ask yourself:

► Did I describe how a plant responds to the Sun?

► Did I include details and sensory words?

► Did I correct all mistakes?

Changes in Ecosystems

Use your textbook to help you fill in the blanks.

What causes an ecosystem to change?

1. Ecosystems do not remain the same. They are

   always _____ .

2. In most ecosystems, change is part of a natural

   _____ .

3. Some changes are long lasting, such as those caused

   by a volcano, hurricane, _____ , or fire.

4. Living things can change a(n) _____ .

   Some living things have a _____
   effect while others can be harmful.

How do people change ecosystems?

5. Some changes that people make to ecosystems are

   helpful, while others can _____ an
   ecosytem.

6. Building roads, homes, and shopping malls affects an

   ecosystem by destroying the _____
   of living things.

7. Some examples of how people harm ecosystems

   are _____ , overpopulation,

   and _____ .

What happens when ecosystems change?

8. Some living things survive changes by changing their

_____ and habits.

9. An individual organism's response to changes is

called _____ .

10. When a species cannot adapt and most of its members

have died, the species is _____ . When

no members are left, the species is _____ .

How can people prevent extinction?

11. Scientists try to keep animals such as panda bears from

becoming _____ by preserving land

where they are _____ .

Critical Thinking

12. Why do you think birds and other small animals might
move to an alligator hole even if an alligator might
eat them?

_____

_____

_____

_____

_____

Changes in Ecosystems

What am I?

Choose a word from the word box below that answers each question.

a. accommodation	**c.** endangered	**e.** overpopulation
b. deforestation	**d.** extinction	**f.** pollution

1. _____ I am the name for a species that only has a small number of members left alive and is in danger of dying out. What am I?

2. _____ I am the result when toxic gas, acid rain, and fertilizers affect an ecosystem. I make the air, land, or water in an ecosystem dirty and unsafe. What am I?

3. _____ I am what happens when a forest is cut down to make room for roads and buildings. What am I?

4. _____ I am the result of more and more living things moving into an ecosystem, taking up more space, and using more natural resources. What am I?

5. _____ I am what happens to an entire species when its last member dies. What am I?

6. _____ I am the ability of some living things to survive changes in an ecosystem by changing their behavior and habits. What am I?

Changes in Ecosystems

Fill in the blanks.

accommodation	helpful	pollution
adapt	hurricane	protecting
harmful	natural	short term

Environments are always changing. An ecosystem

can be changed by _____ events, like

a volcano, drought, or _____ . These

changes can be _____ or long lasting.

Living things can also affect ecosystems. Swarms of

locusts have a _____ effect, but

alligators can have a _____ effect.

People can harm an ecosystem with _____ ,

or help it by _____ its resources.

When organisms' ecosystems are changed, they

survive by changing their habits and behaviors

through _____ . If a species cannot

_____ , its members die out. If a lot of

members die out, the species is endangered. If all of the

members die out, then the species is extinct.

Mail Call

In your textbook, read the letter Clara wrote to the museum scientists. Write the sentence that describes the sudden event that caused the change in the chaparral.

_____

_____

Write the sentences that Clara uses to describe the changes in the chaparral.

1. _____

2. _____

3. _____

4. _____

_____

 Write About It

Predict Read the letter again. Predict what the chaparral will be like next year. What might happen to the environment if there is a drought? Write your predictions in the form of a paragraph.

Name _____ Date _____

Predict

Complete the graphic organizer below. Given the predictions shown, tell what you think will happen.

Prediction	What Will Happen
Another drought will occur in southern California during the summer.	
Another wildfire will occur in the chaparral environment next year because of the lack of rain in the summer.	
Seeds from monkey flower and scarlet larkspur will burn in the wildfires.	
Fields of wildflowers will grow.	
Shrubs and bushes will grow.	

Now, write a paragraph describing what might happen if a drought were to affect the chaparral next year.

_____

_____

_____

_____

© Macmillan/McGraw-Hill

Surviving in Ecosystems

Circle the letter of the best answer.

1. An adaptation that allows an organism to blend into the colors and shapes of its environment is called

    a. accommodation.

    b. hibernation.

    c. mimicry.

    d. camouflage.

2. The roots of a plant grow downward in response to what abiotic factor?

    a. nutrients

    b. gravity

    c. sunlight

    d. soil

3. The response of a plant to a ___ is called tropism.

    a. reaction

    b. growth

    c. dehydration

    d. stimulus

4. Some animals survive the cold winter by ___ , saving energy by remaining completely still for a long period of time.

    a. hibernating

    b. accommodating

    c. stimulating

    d. camouflaging

5. When an entire forest is cut down to build roads or buildings, it is called

    a. accommodation.

    b. adaptation.

    c. deforestation.

    d. deconstruction.

6. Some organisms survive because they can ___ , or look like other, more dangerous organisms in their environment.

    a. respond

    b. mimic

    c. camouflage

    d. accommodate

Name _____ Date _____

Circle the letter of the best answer.

7. A green plant will grow toward the source of this stimulus because the plant needs it in order to make food.

   a. The stimulus is gravity.

   b. The stimulus is water.

   c. The stimulus is light.

   d. The stimulus is noise.

8. Some animals survive a change in their environment by changing their behaviors or habits. This is called

   a. accommodation.

   b. adaptation.

   c. adjustment.

   d. acceptance.

9. A species is ____ when all of its members have died.

   a. environmental

   b. endangered

   c. in the ecosystem

   d. extinct

10. When more organisms move into an ecosystem and use more resources, the result is

   a. overpopulation.

   b. overcrowding.

   c. accommodation.

   d. adjustment.

11. A species is ____ when only a small number of its kind are left.

   a. environmental

   b. endangered

   c. in the ecosystem

   d. extinct

12. Any harmful substance that enters the air, water, or land can cause

   a. overcrowding.

   b. pollution.

   c. extinction.

   d. danger.

13. A trait that helps an organism survive in its environment is

   a. an adaptation.

   b. a reaction.

   c. an accommodation.

   d. a stimulus.

© Macmillan/McGraw-Hill

Lichen: Life on the Rocks
From *Ranger Rick*

Read the Unit Literature feature in your textbook.

Write About It

Response to Literature This article tells you that lichen is not one thing but two. What are the two parts of a lichen? How can a lichen change rocks? Write a summary. Use your own words to explain what this article is about.

_____

_____

_____

_____

_____

_____

_____

_____

_____

_____

_____

_____

Name _____ Date _____

Shaping Earth

Complete the concept map about how Earth is shaped by different events. On each line, write an example of how that term shapes Earth.

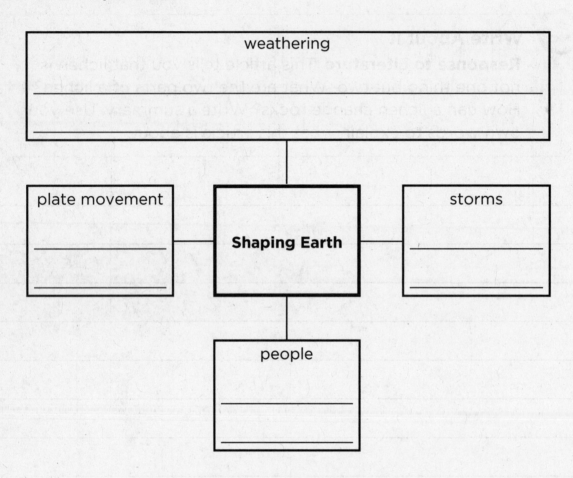

Earth

Use your textbook to help you fill in the blanks.

What does Earth's land look like?

1. A physical feature of the land is called a(n) _____ .

2. The tallest of all landforms are _____ .

3. Vast areas of land without mountains or hills are called

_____ .

4. Valleys and canyons are examples of landforms shaped

by _____ .

5. Mounds, called _____ , form where wind
blows sand.

What does it look like where water meets land?

6. The gently sloping edge of a continent that connects the

shore to the sea is a _____ .

7. Underwater mountains that run through an ocean form

a(n) _____ .

8. Features that look like canyons in the ocean floor are

called _____ .

9. The name of a region of land where water drains into a

river is a _____ .

10. The movement of a river slows down as it nears the ocean, dropping deposits that form triangle-shaped landforms,

called _____ .

What is below Earth's surface?

11. The outermost layer of Earth is made up of rock, called

the _____ .

12. Below the crust lies a layer of rock, called

the _____ .

13. Rock in the mantle can move or slowly flow because of

great pressure and high _____ .

14. The outer core is below the mantle and is made mostly

of melted _____ .

15. The sphere of solid material at Earth's center is called

the _____ .

Critical Thinking

16. Do you think wind, water, or the Earth itself was the last factor to affect the landform where you live?

_____

_____

_____

_____

_____

_____

Earth

Match the correct letter with the description.

a. continental shelf	**d.** drainage basin	**g.** outer core
b. continental slope	**e.** inner core	**h.** plain
c. crust	**f.** mantle	

1. _____ a layer of Earth that is probably made of melted iron

2. _____ a region of land with water that runs into a river

3. _____ solid rock that makes up Earth's outermost layer

4. _____ the area where the edge of a continent falls steeply to the ocean floor

5. _____ the solid layer of Earth that can flow because of great heat and pressure

6. _____ the sphere of solid material at Earth's center

7. _____ the underwater edge of a continent

8. _____ very flat land

Name _____ Date _____

Earth

Fill in the blanks.

continental shelf	inner core	mountains
continental slope	landforms	outer core
crust	mantle	plains

Earth is made of layers. The outermost layer of Earth

is the _____ . Under this layer is the

_____ . Below this layer is melted iron,

called the _____ . At the center of Earth, a

solid mass exists called the _____ .

Earth's largest land masses are the seven continents.

When a continent touches an ocean, it slopes into the ocean

and forms a _____ . As the continent moves

farther into the ocean, it becomes steeper and forms a

second feature, called a _____ .

On land, features called _____

change the flatness of the land. These landforms include

_____ and _____ .

The crust contains all of the features that rise from Earth's

surface and form the ocean floor.

The Moving Crust

Use your textbook to help you fill in the blanks.

What distorts Earth's crust?

1. The large moving sections that make up Earth's crust and

   upper mantle are called _____ .

2. When plates slowly ram into each other, they can form

   bended rock layers, called _____ .

3. A crack in Earth's crust along which movement takes place

   is called a(n) _____ .

4. Tall landforms caused by folding or faulting are

   called _____ .

What causes earthquakes?

5. If the rock in Earth's crust suddenly shakes, a(n)

   _____ occurs.

6. Underwater earthquakes can cause huge ocean waves,

   called _____ .

How do scientists study earthquakes?

7. The source of an earthquake creates _____

   _____ that travel outward.

8. A tool that graphs seismic waves as wavy lines is called

   a(n) _____ .

9. Seismic waves travel at different _____ along

Earth's _____ and _____
Earth's interior.

What is a volcano?

10. A mountain that forms around an opening in Earth's crust

is a(n) _____ .

11. Most volcanoes form near _____ edges.

12. A volcano is produced by melted rock, called

_____ , which erupts onto the surface as

_____ .

13. Some volcanoes form over thin places in Earth's crust,

called _____ .

14. In the Pacific Ocean, an example of volcanoes that formed

over a hot spot is the _____ Islands.

Critical Thinking

15. Do you think there are earthquakes in the
Hawaiian Islands? Explain.

_____

_____

_____

_____

_____

© Macmillan/McGraw-Hill

The Moving Crust

Fill in the blanks.

earthquake	fold	plateau	seismograph
fault	mountain	seismic wave	volcano

1. A crack in Earth's crust along which movement takes place

   is called a(n) _____ .

2. A tall landform that rises to a peak is called

   a(n) _____ .

3. A tool that measures the waves from an earthquake is

   a(n) _____ .

4. A(n) _____ is a bend in rock layers.

5. A(n) _____ is a high landform with a
   flat top.

6. A(n) _____ moves outward from the
   source of an earthquake.

7. An opening in Earth's crust through which gases and

   melted rock pass is a(n) _____ .

8. When the rock along a fault moves suddenly, a(n)

   _____ occurs.

The Moving Crust

Fill in the blanks.

earthquake	mountains	source
faults	plateaus	volcanoes
folds	seismic waves	

Many changes in Earth's crust are caused by the

movement of large sections of the crust, called plates.

Sometimes when plates meet, rock layers bend and

form _____ . Openings in Earth's crust,

called _____ , also occur where plates

meet. As plates get pushed and pulled cracks form

called _____ , and the two sides move

in opposite directions. When the plates move slowly,

_____ and _____ can form.

If the plates move quickly, a(n) _____

can occur. Earthquakes send out _____ that

move in all directions from the _____ .

A seismograph is a tool that scientists use to record and

measure the seismic waves of earthquakes.

Meet Ro Kinzler

Read the passage in your textbook about Ro Kinzler. Then list the places that Ro Kinzler and some of the other scientists have traveled.

1. _____

2. _____

Where do Ro and the other scientists perform experiments to test their findings?

_____

_____

What are some of the things they have done with their samples and observations?

1. _____

_____

2. _____

_____

Write About It

Cause and Effect Read the article with a partner. Fill out a cause-and-effect chart to record why Ro visits volcanoes and collects lava samples. Tell what happens as a result of her work.

Name _____ Date _____

Cause and Effect

Use the answers to the questions to complete the cause-and-effect chart.

Cause	Effect
Ro travels to the _____ .	She collects active lava samples to study.
Ro goes to the ocean floor.	She creates _____ of the ocean floor based on careful observations of rock formations.

Guidelines—What to write in the chart:

▶ Look for why something happened. This is the cause.

▶ Look for what happened as a result. This is the effect.

Cause	Effect
Ro would go just about anywhere.	She finds out more about _____ .
Ro does experiments on lava.	She finds out how lava _____ .
Ro goes to the _____ .	She sees underwater volcanoes.
Ro observes underwater _____ .	She creates maps of the ocean floor.

Weathering and Erosion

Use your textbook to help you fill in the blanks.

What is weathering?

1. The slow process that breaks rocks into smaller pieces is

   called _____ .

2. A rock is broken apart by _____ weathering if the rock type does not change.

3. If a rock contains iron, air and water can react with the iron

   through _____ and form rust.

What is erosion?

4. The weathering and removal of rock from one place to

   another is called _____ .

5. Erosion can be caused by glaciers, wind, moving water,

   and _____ .

6. When the Colorado River eroded the land around the river

   in Arizona, the _____ was formed.

How do glaciers shape the land?

7. Glaciers form in very cold places as thick _____ of ice.

8. As the weight of the overlying ice increases, the glacier begins to

   _____ .

9. Deposits left behind by a glacier are called

   _____ and _____ .

10. The mounds that form where till builds up are

called _____ .

How do people shape the land?

11. Most processes change land slowly, but people can

make _____ changes.

12. People change the land by _____
it to get minerals, metals, or fuels.

Critical Thinking

13. Which do you think changes the land more: frozen water or
flowing water?

_____

_____

_____

_____

_____

_____

_____

_____

_____

_____

_____

_____

_____

© Macmillan/McGraw-Hill

Weathering and Erosion

Secret Word

Read each clue. Write the answer in the correct squares in the puzzle. Then, figure out what the secret word is, and fill in the rest of the letters.

debris	glacial till	physical	weathering
erosion	moraine	terminus	

Across

1. rocks or gravel left by a glacier

2. a slow process that breaks rocks into smaller pieces

3. the carrying away of weathered pieces of rock

4. weathering that breaks down rock without changing the rock type

5. an unsorted mixture of debris dropped by a glacier

6. the bottom end of a glacier

7. features that form where glacial till builds up

Write the secret word that is running down the puzzle.

_____

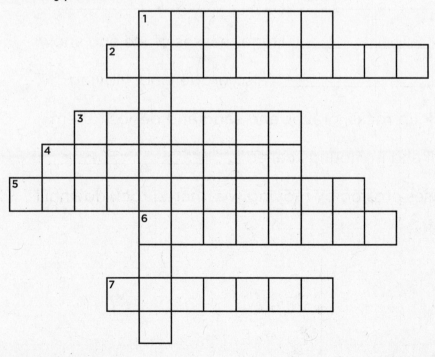

Name _____ Date _____

Weathering and Erosion

Fill in the blanks.

chemical weathering	moraines	wind
glaciers	physical weathering	
gravity	water	

  Rocks are constantly broken down by two processes,

called weathering and erosion. Rocks can be broken down

into smaller pieces without changing the type of rock

through _____ . Minerals in rocks can be

changed to other minerals through _____ .

Erosion moves weathered rock from one place to another

through _____ , _____ ,

and _____ . Huge masses of ice and snow,

called _____ , also erode land. Moving

glaciers pick up rocks, gravel, and sand and deposit them

as glacial till and in mounds called _____ .

Gravity causes erosion by moving weathered rock downhill.

Land Over Time

> ### Write About It
> **Expository Writing** Write a paragraph that summarizes "Land Over Time." Include the main idea and the most important details.

Getting Ideas

Think about what you read in "Land Over Time."
Then fill in the summary chart.

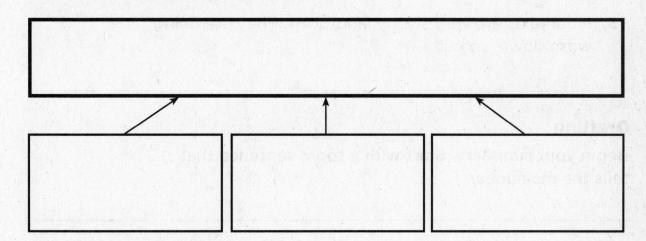

Planning and Organizing

Brandon wrote three sentences. Write "MI" next to the sentence that tells the main idea. Write "D" next to each sentence that tells a detail.

1. _____ Water freezes when the temperature drops below 32°F.

2. _____ When water freezes, it expands, and the cracks in rocks get bigger.

3. _____ Weather changes land over a long period of time.

Revising and Proofreading

Here are some sentences that Brandon wrote. He wants to join them together. Read the sentences. Then look at the pair of words. Circle the word that best fits to join the sentences. Then write the new sentence on the line. Put a comma before the word.

1. Rocks have small cracks. The rain fills the cracks.   and but

    _____

2. Roots get thicker. The cracks widen over time.   but so

    _____

3. Mountains are very sturdy landforms. They are being worn down.   so but

    _____

Drafting

Begin your summary. Start with a topic sentence that tells the main idea.

_____

_____

Now write your summary. Use a separate piece of paper. Start with the topic sentence you wrote above. Include only important facts and details from "Land Over Time." Put them in your own words.

Now revise and proofread your writing. Ask yourself:

► Did I tell only the most important information?

► Did I draw a conclusion based on the information presented?

► Did I correct all mistakes?

Changes Caused by the Weather

Use your textbook to help you fill in the blanks.

How do floods and fires change the land?

1. An overflow of water onto land that is normally dry is

   called a(n) _____ .

2. Flood waters usually come from heavy _____ or quickly melting snow.

3. A flood can help the land by depositing rich _____ on it.

4. A fire can be caused by _____ or other natural sources.

5. Plants and animals may lose their _____ after a fire.

How do storms change the land?

6. The storm caused by a column of air that spins rapidly is

   called a(n) _____ .

7. Usually, tornadoes spin off of severe _____ .

8. Tornadoes are common in the _____ region of the United States.

9. A swirling system of winds, huge walls of clouds, and

   pounding rains is called a(n) _____ .

10. Hurricanes form over warm _____ .

11. Hurricanes can last for many days and stretch for hundreds

of _____ .

12. Hurricanes are becoming more and more common in some

places, possibly because of higher _____ .

How do landslides change the land?

13. Rocks and water-soaked soil move quickly down a hillside

during a(n) _____ .

14. Tons of snow and ice move suddenly down a mountain

during a(n) _____ .

15. A cause of landslides is _____ , which
pulls rocks from high places to low places.

Critical Thinking

16. Which do you think causes more damage to Earth: too
much water or not enough water?

_____

_____

_____

_____

_____

_____

_____

Changes Caused by the Weather

Match the correct letter with the description.

a. avalanche	**c.** flood	**e.** landslide	**g.** tornado
b. fire	**d.** hurricane	**f.** gravity	**h.** lightning

1. _____ a column of air that spins rapidly

2. _____ a swirling system of winds, huge clouds, and pounding rains

3. _____ can be fueled by dry plants and spread by wind

4. _____ an overflow of water onto land that is normally dry

5. _____ the force that pulls rocks and other objects from a high place to a low place

6. _____ sliding ground caused when a hillside is soaked with rainwater

7. _____ the sudden movement of tons of snow and ice down a hill or mountain

8. _____ can cause natural forest fires

Name _____ Date _____

Changes Caused by the Weather

Fill in the blanks.

| fire | hurricane | soil | tornado |
| flood | landslide | thunderstorms | |

Violent weather, or storms, can cause quick changes

in the land. A column of air that spins rapidly is called a(n)

_____ . Tornadoes usually form with severe

_____ . A larger and longer-lasting storm

with heavy rain and swirling winds is called

a(n) _____ .

When heavy rain soaks the ground on a hillside, a(n)

_____ can occur. Heavy rains or melting

snow can also cover land that is usually dry, causing a(n)

_____ . Floods destroy property, but

they also leave rich _____ on the land.

When there is not enough rain, plants become dry, and

_____ can damage the land. Fires can be

started by natural events, such as lightning, and the actions

of careless people.

© Macmillan/McGraw-Hill

Shaping Earth

Circle the letter of the best choice.

1. Mountains, plains, and plateaus are examples of
   a. avalanches.
   b. landforms.
   c. landslides.
   d. ridges.

2. Which underwater feature is similar to a canyon?
   a. the continental shelf
   b. the continental slope
   c. an ocean ridge
   d. a trench

3. What is the outermost layer of Earth?
   a. the crust
   b. the inner core
   c. the mantle
   d. the outer core

4. Which of the following is NOT a way in which mountains form?
   a. folding rock
   b. landslides moving large amounts of land
   c. rocks moving along a fault
   d. volcanoes erupting

5. What moves in all directions from the source of an earthquake?
   a. faults
   b. folds
   c. seismic waves
   d. seismographs

6. Which statement is true about volcanoes?
   a. Lava and ash come out of an erupting volcano.
   b. Volcanoes erupt only on land.
   c. Volcanoes form over cold spots.
   d. Volcanoes only form at the point where two plates meet.

7. A large, slow-moving section of Earth's crust and upper mantle is called a
   a. continent.
   b. fault.
   c. fold.
   d. plate.

Circle the letter of the best choice.

8. Acid rain changes the minerals in rocks to other minerals. What is this process called?

   a. chemical weathering

   b. deposition

   c. erosion

   d. physical weathering

9. Which of the following was formed by erosion?

   a. the Grand Canyon

   b. the Great Plains

   c. the Hawaiian Islands

   d. the Mississippi River

10. What is a large, slow-moving buildup of snow and ice?

   a. an avalanche

   b. a flood

   c. a glacier

   d. a landslide

11. The unsorted mixture of rocks dropped by a glacier along its end or sides is called

   a. debris.

   b. glacial till.

   c. a moraine.

   d. a terminus.

12. What happens when water covers land that is usually dry?

   a. a fire

   b. a flood

   c. a hurricane

   d. a tornado

13. What kind of storm forms over an ocean?

   a. a flood

   b. a hurricane

   c. a dust storm

   d. a tornado

Saving Earth's Resources

Use your textbook to help you fill in the blanks.

```
┌─────────────────────────────┐
│      Earth's Resources      │
└─────────────────────────────┘
```

Rocks	**Soil**	**Water**	**Minerals**
are made of	is needed for	is needed by	are found in
_____	_____	_____	_____
and are classified by how they are	and is made of	and freshwater is found mostly in	and is used to make
_____	_____	_____	_____
.	_____	_____	_____
	_____		_____
	.		_____
			_____ .

Name _____ Date _____

Minerals and Rocks

Use your textbook to help you fill in the blanks.

What is a mineral?

1. Nonliving, natural substances that make up rock are

   called _____ .

2. Minerals can be identified by their _____ :

   color, _____ , luster, and streak.

3. Minerals are the building blocks of _____ .

What are igneous and sedimentary rocks?

4. Minerals offer clues about how a rock _____ .

5. A rock that froms from melted rock is

   _____ rock.

6. Melted rock, called _____ , cools and
   hardens to form igneous rock.

7. Igneous rocks are classified by how quickly they

   _____ and the size of the mineral
   grains that form.

8. Layers of substances are pressed and cemented together

   to form _____ rock.

9. Scientists compare the position of layered rocks to find

   a rock's _____ age.

10. A rock layer and any _____ in that
    rock have the same relative age.

What are metamorphic rocks?

11. Heat and pressure form _____ rocks.

12. Metamorphic rocks can form from _____

and _____ rocks, or from other
metamorphic rocks.

13. Metamorphic rocks are different than their original form

because their _____ change.

14. Rocks change from one form to another in the

_____ .

How do we use rocks?

15. People use rocks and minerals as _____ .

16. Rocks and minerals are used for _____
schools and other structures, and to make steel, aluminum

products, silicon chips, and _____ .

Critical Thinking

17. Which type of rock (metamorphic, sedimentary, or
igneous) do you think is most useful?

_____

_____

_____

_____

_____

_____

Name _____ Date _____

Minerals and Rocks

Match the correct letter with the description.

a. igneous rock	**d.** properties	**g.** rock cycle
b. metamorphic rock	**e.** relative age	**h.** sedimentary rock
c. mineral	**f.** resource	

1. _____ a natural, nonliving substance that occurs in rocks

2. _____ results from melted rock that cools and hardens

3. _____ results from sediment that is pressed and cemented together

4. _____ the age of a rock that is found by comparing the position of the rock to other rock layers

5. _____ formed by heat and pressure

6. _____ the name of the process by which rocks change from one form to another

7. _____ any material found on Earth that can be used by people

8. _____ characteristics like color, hardness, luster, and streak

Minerals and Rocks

Fill in the blanks.

fossils	nonliving	rock cycle
igneous rock	properties	sedimentary rock
metamorphic rock	relative ages	

Minerals come from Earth's crust. They are found as

_____ solids. Many minerals and rocks have

_____ that make them useful to people.

Scientists classify rocks in three main groups. Melted

rock that cools and hardens forms _____ .

Sediments that have been pressed and cemented together

over thousands of years form _____ , which

can also contain _____ . Scientists can find

the _____ of rocks and fossils by looking

at their position in layers of rock. Rocks formed by heat

and pressure are called _____ . All types

of rocks move through the _____ . This is

a neverending process by which Earth's rocks change from

one form to another.

Name _____ Date _____

Soil

Use your textbook to help you fill in the blanks.

What is soil made of?

1. If you look at soil with a hand lens, you will find small

   pieces of rocks, minerals, and _____ .

2. You might not see them, but soil is made up of

   _____ , air, and _____

   as well.

3. When plants and animals die, _____ and
   fungi decompose them.

4. All the layers of soil, from Earth's surface down to the

   bedrock, are shown in a(n) _____ .

5. The different layers of a soil profile are called _____

   and include topsoil, _____ , and weathered

   _____ .

What are some properties of soil?

6. The property of soil that refers to the size of soil particles

   is called _____ . Clay is fine with small

   particles and sandy soil is _____ with
   large particles.

7. Other properties of soil are color and _____ ,

   which indicates how easily _____ passes
   through soil.

8. The spaces between the particles in soil, called pore

   spaces, determine how _____ the soil is.

Why is soil type important?

9. Soil that is permeable to air and _____ will
   allow living things to survive.

10. If the soil does not hold enough water, crop plants can

    _____ . If the soil holds too much water,

    crop plants can _____ .

Critical Thinking

11. What do you think pedologists, scientists who study soil,
    can learn by picking up a handful of soil?

    _____

    _____

    _____

    _____

    _____

    _____

    _____

    _____

    _____

Soil

What am I?

Choose a word from the word box below that answers each question.

a. horizon	**c.** permeability	**e.** porous	**g.** subsoil
b. humus	**d.** pore spaces	**f.** soil profile	**h.** topsoil

1. _____ I am the decayed plant and animal matter in soil. What am I?

2. _____ I am a section of soil that includes all layers from the topsoil to the bedrock. What am I?

3. _____ I am the name for a layer of soil. What am I?

4. _____ I am the name for the surface layer of soil. What am I?

5. _____ I am the layer of soil under the top layer of soil. What am I?

6. _____ I am the spaces between the particles that make up soil. What am I?

7. _____ I am a characteristic of soil that depends on the size and number of its pore spaces. What am I?

8. _____ I am any material that lets water pass through it. What am I?

Soil

Fill in the blanks.

horizons	pore spaces	resource
humus	porous	soil profile
minerals	properties	weathered rock

Farmers use soil to grow plants that are used for

food and other products. Soil is an important natural

_____ . It is a mixture of small pieces

of rock and decayed plant and animal matter, called

_____ . Soil contains _____

that can be filled with air or water, meaning that it

is _____ .

Soil forms in layers called _____ . Each

layer of soil has its own _____ that make

up a(n) _____ . The top layer, or topsoil, is

rich with humus and _____ . The next layer

is called subsoil. The bottom layer is made of broken bits

of bedrock or _____ . Soil is different from

place to place because the minerals and rocks are different.

Resources from the Past

Use your textbook to help you fill in the blanks.

What are fossils?

1. Evidence of a living thing from long ago is preserved as

   a(n) _____ .

2. Many fossils are found preserved in _____
   rock.

3. Fossils also form in spaces left in rock by dead organisms.

   These spaces are called _____ . Minerals

   fill the spaces, forming fossils called _____ .

4. Some fossils are formed when an organism such
   as a leaf is pressed into a soft surface, leaving a(n)

   _____ .

5. An entire organism can be preserved, such as an insect

   trapped in _____ or a mammoth trapped

   in _____ .

How do we study fossils?

6. Fossils tell how Earth's land, _____ , and
   living things have changed over time.

7. Fossils and layers of rock are evidence of changes that

   took place over long spans of _____ time.

What are fossil fuels?

8. Something that can supply _____ is a fuel.

9. Fuel that is made from the remains of ancient living things

   is a(n) _____ .

10. Fossil fuels form from the remains of buried _____

    and _____ .

11. Fossil fuels are _____ resources because
    they cannot be replaced easily.

What can we use instead of fossil fuels?

12. Air, water, plants, and animals are _____

    resources because they can be replaced in _____ .

13. Renewable energy sources include solar energy,

    _____ that harness the wind, and energy

    from _____ inside Earth.

Critical Thinking

14. Do you think you could go to school without using
    fossil fuels?

    _____

    _____

    _____

    _____

    _____

Resources from the Past

Choose a word from the word box below that correctly fills in the blank.

amber	fossil	imprint	nonrenewable
cast	fossil fuel	mold	renewable

1. An energy source that cannot be replaced easily is a(n)

   _____ resource.

2. Fuel made from the remains of ancient living things is

   called _____ .

3. Minerals that fill the space left in a rock by a decayed

   organism form a fossil called a(n) _____ .

4. An energy source that nature replaces quickly is a(n)

   _____ resource.

5. The space in a rock left by the remains of an organism is

   a(n) _____ .

6. Sap from a tree that hardens into a hard, yellow material

   forms _____ .

7. The preserved evidence of a living thing from long ago is

   a(n) _____ .

8. The fossil formed by an organism that has been pressed

   into a soft surface is a(n) _____ .

© Macmillan/McGraw-Hill

Resources from the Past

Fill in the blanks.

casts	fossil fuels	imprints	nonrenewable
energy	fossils	molds	renewable

Earth's land, water, climate, and living things on Earth have undergone many changes over time. Scientists study remains from long ago, called _____ , to learn about life long ago. They study _____ that form in spaces, _____ that fill those spaces, _____ preserved in soft surfaces, and fossils found in amber and ice. The remains of living things buried for millions of years are also a(n) _____ source. Coal, oil, and natural gas are _____ that are burned to release their stored energy.

Fossil fuels are _____ energy sources that take millions of years to form and cannot be replaced easily. Solar energy and energy from moving water and wind are _____ sources of energy. Scientists are working to find ways to use renewable energy sources before we use up all of our fossil fuels.

Water

Use your textbook to help you fill in the blanks.

Where is Earth's water found?

1. Most of Earth's fresh water is found in solid form as

   _____ and _____ .

2. Streams, rivers, and lakes hold the rest of Earth's fresh

   water in _____ form.

3. Rain soaks into the soil and collects there

   as _____ .

4. The place where water drains downhill into one river is

   called a(n) _____ .

How is fresh water supplied?

5. Water from lakes and rivers is sometimes stored in a(n)

   _____ so that people will have a year-
   round water supply.

6. Rain or melted snow that flows over land is called

   _____ . It may carry _____

   or harmful _____ .

7. Digging a well is a common way to get _____ .

© Macmillan/McGraw-Hill

8. Most wells have a(n) _____ to bring groundwater to the surface.

9. Impurities are removed from water at a(n) _____ plant to make it safe to use.

How do we use water?

10. People use fresh water for _____ .

11. We use water to generate _____ .

12. Farmers use _____ to bring water through pipes and ditches to their crops.

13. People also use water for fun activities, including

_____ , boating, and _____ .

Critical Thinking

14. How do you think the water in a watershed is used?

_____

_____

_____

_____

_____

_____

_____

_____

Name _____ Date _____

Water

Match the correct word with the description.

a. groundwater	**c.** reservoir	**e.** seawater	**g.** watershed
b. irrigation	**d.** runoff	**f.** soil water	**h.** well

1. _____ rain or melted snow that flows over land and collects in streams or brooks

2. _____ a region where water from different sources drains into one river

3. _____ rain that soaks into the ground and is used by plants

4. _____ water that is stored in spaces in underground rocks

5. _____ a deep hole dug in the ground to reach groundwater

6. _____ an artificial lake built by people to hold fresh water

7. _____ a method used by farmers to bring water to their crops through pipes or ditches

8. _____ another name for ocean water that contains too much salt for people or plants

Water

Fill in the blanks.

fresh water	liquid	salt water	wells
irrigation	reservoirs	solid	

Oceans cover about three-fourths of Earth's surface.

Ocean water is not useful to most living things, including

people, because it is _____ . Most living

things need _____ to survive. Fresh water

is found as a(n) _____ in ice caps and

glaciers. It is found as a(n) _____ in streams,

rivers, and lakes.

Fresh water can be pumped to the surface through

_____ or used by farmers for

_____ . Fresh water from lakes and rivers

is stored in _____ for year-round use in

nearby communities. Harmful materials are removed at water

treatment plants before the water is sent to homes, schools,

and hospitals.

© Macmillan/McGraw-Hill

Name _____ Date _____

Saving Water

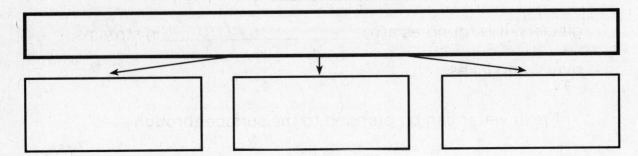

Write About It

Persuasive Writing Write a letter to the editor of your local newspaper. Your letter should inform people about the need to keep the groundwater clean. Include facts and details to make your letter persuasive.

Getting Ideas

Research how groundwater becomes polluted. Then use the chart below to help you organize your ideas for keeping groundwater clean. Write your opinion in the top box. Write three reasons in the bottom boxes.

Planning and Organizing

Isabella wrote three sentences. Write "O" next to the sentence that tells her opinion. Write "R" next to each sentence that tells a convincing reason to keep groundwater clean.

1. _____ Water animals need clean places to live and clean water to drink.

2. _____ I strongly believe that we must do more to keep our water clean.

3. _____ People need clean drinking water to live.

Revising and Proofreading

Here are some sentences that Isabella wrote. There are some errors in punctuation. Proofread the sentences. Find the five errors. Insert the correct punctuation mark, or delete the punctuation mark if it is incorrect.

Dear Editor:

Every day I walk by the pond and there are very few water lilies blooming. Dr Mcgregor says that the groundwater is polluting the pond. It is because of the factory. There are leaks in the tanks and I believe this is a serious problem. Chemicals leak into Earth. This causes the pollution. What is the paper's position on this serious problem. Isnt it time you spoke up?

Drafting

Begin your letter. Start by stating the reason for your letter.

_____

_____

Now write your letter. Use a separate piece of paper. Start with the sentence you wrote above. Write about the importance of keeping our groundwater clean. Include convincing reasons. Make sure you tell the reader what actions could be taken to solve this problem.

Now revise and proofread your writing. Ask yourself:

▶ Did I clearly state my opinion about this topic?

▶ Did I use convincing reasons and arguments?

▶ Did I correct all mistakes?

© Macmillan/McGraw-Hill

Name _____ Date _____

Pollution and Conservation

Use your textbook to help you fill in the blanks.

What is pollution?

1. All of the living and nonliving things in an area make up

   the _____ .

2. Dangerous or harmful materials cause _____
   when they are added to the environment.

3. Pollution can be caused by _____ events

   or human _____ .

4. Waste gases in the air cause _____ .

   They create _____ or combine

   with water droplets to create _____ .

5. Oil spills, fertilizers, and pesticides in oceans and streams

   can cause _____ pollution.

6. If trash in landfills is not stored in the right way, it can

   cause _____ pollution.

How can we protect the soil and water?

7. One way to protect the environment is to use

   _____ wisely by practicing _____ .

© Macmillan/McGraw-Hill

8. Soil conservation includes activities that keep soil healthy

for growing _____ .

9. Towns and cities conserve their water supply by cleaning

wastewater at _____ treatment plants.

10. People practice _____ conservation
whenever they use water wisely.

What are the 3 *R*s?

11. The 3 *R*s are main ways to _____
resources.

12. People can conserve resources and _____
waste by using less of something.

13. People can _____ things instead of
throwing them away after one use.

14. People conserve when they _____ by
making something new from used materials.

Critical Thinking

15. Do you think that reducing pollution and practicing
conservation work better in big cities or small cities?

_____

_____

_____

_____

_____

_____

Name _____ Date _____

Pollution and Conservation

Use the clues below to help you find the words hidden in the puzzle.

1. All of the living and nonliving things in a certain area make

   up its _____ .

2. Harmful materials found in the environment cause

   _____ .

3. Waste gases combined with water droplets form

   _____ .

4. Using resources wisely is

   _____ .

5. A fertilizer made from decaying table scraps and dead plants is

   _____ .

6. The 3 *R*s of conservation

   are _____ ,

   _____ , and

   _____ .

N	O	I	T	A	V	R	E	S	N	O	C
R	S	G	C	A	C	I	D	R	A	I	N
E	N	V	I	R	O	N	M	E	N	T	A
D	K	Y	D	J	M	V	U	C	F	G	U
U	M	F	R	H	P	S	D	Y	A	J	P
C	S	M	A	G	O	G	A	C	M	V	Y
E	B	R	E	U	S	E	D	L	N	S	W
W	A	S	N	D	T	D	P	E	I	F	D
G	P	N	O	I	T	U	L	L	O	P	M
R	E	C	Y	C	L	E	Z	E	A	A	Y

Pollution and Conservation

Fill in the blanks.

acid rain	fertilizers	smog
ashes	pollution	smoke
conservation	resources	water

Living things get what they need from their environment.

Living things need clean air, clean _____ ,

and healthy land to survive. Harmful materials enter the

environment and cause _____ . Natural

events, such as forest fires and erupting volcanoes, can

pollute the environment with _____ and

_____ . Waste gases from burning fossil

fuels can produce _____ . These gases may

also combine with water droplets in the air and produce

_____ .

Pesticides and _____ pollute both

the water and the land. People can help control pollution

by using _____ wisely and practicing

_____ . People must practice the 3 Rs of

conservation: reduce, reuse, and recycle.

Name _____ Date _____

Saving the Soil

Read the passage in your textbook. As you read, write down the topic of each paragraph. Also, pay attention to the supporting details about saving the soil.

Topic sentences:

1. _____

2. _____

3. _____

4. _____

Write About It

Main Idea and Details

1. Why do farmers need to protect the soil?
2. What are some ways that farmers protect the soil? List the advantages and disadvantages of these methods.

_____

_____

_____

_____

_____

_____

Use the Main Idea and Details graphic organizer to record information about what farmers do to save the soil.

Main Idea	Details
Contour plowing is helpful.	It is used on a _____ to stop _____ from flowing downhill and carrying away soil and _____ .
Contour plowing is not easy.	It takes more _____ and uses more _____ than straight plowing.
Some farmers do not plow their fields after a harvest.	Instead, they place _____ in holes in the ground.
After the harvest, some farmers plant cover crops, like _____ .	This adds _____ and _____ the ground.
Some farmers use _____ planting instead of plowing.	They dig_____ and drop in _____ . Farmers may have to use _____ to kill weeds that plowing would have removed.

Saving Earth's Resources

Circle the letter of the best answer.

1. What is a mineral?

   a. a nonliving substance formed in nature

   b. a nonliving substance made only in factories

   c. a living substance formed in nature

   d. a living substance found only in fossils

2. The type of rock formed when melted rock cools and hardens is

   a. magma.

   b. sedimentary.

   c. igneous.

   d. metamorphic.

3. The neverending process by which rocks change from one form to another is

   a. a type of life cycle.

   b. the rock cycle.

   c. the cooling of magma.

   d. the relative age of a rock.

4. What does a soil profile show?

   a. the color of the soil under the surface

   b. the layers of soil called horizons

   c. the texture of the soil under the surface

   d. the pore spaces in the soil under the surface

5. The rate at which water flows through soil is known as the soil's

   a. pore space.

   b. texture.

   c. permeability.

   d. profile.

6. What is the name for the preserved evidence of once-living things?

   a. molds

   b. fossils

   c. casts

   d. imprints

Circle the letter of the best answer.

7. What types of resources are fossil fuels such as coal, oil, and natural gas?

   a. nonrenewable

   b. renewable

   c. preserved

   d. unlimited

8. A lake built by people to hold water is a

   a. treatment plant.

   b. reservoir.

   c. watershed.

   d. well.

9. Bringing water to a farmer's fields through pipes and ditches

   a. is called a reservoir.

   b. provides transportation.

   c. is called irrigation.

   d. produces electricity.

10. All of the living and nonliving things in an area make up

   a. the environment.

   b. a landfill.

   c. harmful pollution.

   d. a community.

11. What is pollution?

   a. a natural resource in the environment

   b. harmful material in the environment

   c. anything that happens in the environment

   d. a manmade resource in the environment

12. Protecting resources and using them wisely is known as

   a. conservation.

   b. pollution.

   c. crop rotation.

   d. irrigation.

13. Making a new and different product from old materials is called

   a. reusing.

   b. recycling.

   c. reducing.

   d. remaking.

© Macmillan/McGraw-Hill

Tornado Tears Through Midwest

From *Time for Kids*

Read the Unit Literature feature in your textbook.

Write About It

Response to Literature What would happen if a tornado struck your community? Write a fictional story. Describe how your community would stay safe. How would it rebuild after the disaster?

_____

_____

_____

_____

_____

_____

_____

_____

_____

_____

_____

Weather and Climate

Fill in the blanks in the graphic organizer below with facts you have learned from the chapter on weather and climate.

Weather Event	Cause	Effect

Severe Thunderstorm
A cold air mass moves in quickly and _____ a(n) _____ .

Heavy precipitation falls to the ground. _____ can be seen and _____ can be heard.

Light Rain
A warm air mass _____ a cold air mass.

Light precipitation falls, but there is no _____ or _____ .

Hurricane
A wide storm forms over _____ water.

A very large storm forms with very fast _____ and heavy _____ .

Rainy Weather (that lasts for days)
Two air masses are not moving into one another, creating a(n) _____ .

_____ falls for several days.

Air and Weather

Use your textbook to help you fill in the blanks.

What is in the air?

1. The blanket of air surrounding Earth is called

   the _____ .

2. The atmosphere is made up mostly of _____

   and _____ .

3. The four layers of Earth's atmosphere, from

   lowest to highest, are _____ ,

   _____ , mesosphere, and thermosphere.

What is weather?

4. The condition of the atmosphere at a given time and

   place is called _____ .

5. When you measure how hot or cold something is, you

   measure its _____ . Three factors that
   affect air temperature include time of day or night,

   _____ , and closeness to oceans.

6. Warm air particles are _____ dense, or
   packed together, than cold air.

7. A measure of the amount of moisture in the air

   is _____ .

8. As air cools, the air pressure _____ .

© Macmillan/McGraw-Hill

9. Moving air is called _____ .

10. Air moves when the Sun's _____ heats the air.

11. Any form of water that falls from the clouds is

_____ . The term includes rain, snow,

sleet, and _____ .

How can you measure weather?

12. Scientists collect data from a _____ .

13. To measure rainfall, scientists collect rain in a tube

called a(n) _____ .

14. A tool used to measure air pressure is called

a(n) _____ .

Critical Thinking

15. What weather tools do you think are used in the desert, the humid tropical rain forest, and the frozen tundra?

_____

_____

_____

_____

_____

Name _____ Date _____

Air and Weather

What am I?

Choose a word from the word box below that answers each question.

a. air pressure	**d.** rain gauge	**g.** wind
b. barometer	**e.** temperature	**h.** wind vane
c. humidity	**f.** thermometer	

1. _____ I am the weight of the air above you. What am I?

2. _____ I am moving air. What am I?

3. _____ I can tell you how hot or cold the air is. What am I?

4. _____ I can tell you how much it rained. What am I?

5. _____ I can tell you what the air pressure is. What am I?

6. _____ I point in the direction of the wind. What am I?

7. _____ I am the amount of moisture in the air. What am I?

8. _____ I am a measure of how hot or cold something is. What am I?

Air and Weather

Fill in the blanks.

air pressure	lowest	thermometer
barometer	particles	troposphere
higher	temperature	weight

Weather is the condition of Earth's atmosphere at any

given time and place. All weather takes place in the

_____ level of the atmosphere, called

the _____ . A measure of how hot or cold

the air is, or _____ , is found with a(n)

_____ . Cool air particles are packed more

closely than warm air _____ .

A measure of the _____ of the air

pushing down on an area is called _____ .

It is measured with a(n) _____ . Cool

air has a(n) _____ air pressure than

warm air. A difference in air pressures causes the

movement of air, or wind.

Watching Spring Weather

> **Write About It**
>
> **Expository Writing** Observe the weather in your area every day for two weeks. Record the temperature, air pressure, precipitation, clouds, and wind speed. Write a newspaper article about the changes you observed.

Getting Ideas

Use the information you recorded to fill out the chart below. Under main idea, write an important idea about the weather. Then write facts and details that support your main idea.

Main Idea	Details

Planning and Organizing

Here are some sentences Zack wrote about the weather in his area. Write "MI" if the sentence tells the main idea. Write "D" if it tells a detail.

1. _____ At first, the temperature was in the 70s.

2. _____ The weather has changed a lot during the last two weeks.

3. _____ There wasn't a cloud in the sky.

© Macmillan/McGraw-Hill

Revising and Proofreading

Here are some sentences Zack wrote. Combine each pair of sentences. Use the transition word in parentheses.

1. There has been a big threat of forest fires. It hasn't rained in two weeks. (because)

_____

_____

2. Brush fires start. Leaves and grass dry out from the wind. (when)

_____

_____

3. The weather report said to expect thunderstorms. There is a warm air mass moving through our region. (because)

_____

_____

Drafting

Write a sentence to begin your article about weather in your area. Tell your main idea about how it changed.

_____

_____

Now write your article. Use a separate piece of paper. Remember to include specific details such as the amount of rainfall.

Now revise and proofread your writing. Ask yourself:

▶ Did I tell a main idea about the weather?

▶ Did I include facts and details to back up this idea?

▶ Did I correct all mistakes?

The Water Cycle

Use your textbook to help you fill in the blanks.

Why does water change state?

1. Water moves from Earth's surface into the _____ .

2. Water changes _____ as it moves.

3. Water in the gas state is called _____ .

4. The process during which a liquid slowly changes to

   a gas is called _____ . Heat from the

   _____ causes ocean water to evaporate.

5. The process during which a gas changes to a liquid

   is called _____ . When the air cools,
   water vapor condenses on objects; for example,

   _____ forms on grass.

Where does water go?

6. Earth's water is constantly changing state by moving

   through the _____ .

7. When water vapor rises, it cools and _____ ;
   this forms clouds.

8. When water evaporates from the leaves of plants, it is

   called _____ .

9. Rain, snow, sleet, and hail are different forms

   of _____ .

© Macmillan/McGraw-Hill

What are some types of clouds?

10. Low, layered clouds are called _____ clouds.

11. White, puffy _____ clouds can become

    thick and dark _____ clouds that produce
    precipitation.

12. Thin, wispy clouds high in the sky are called

    _____ clouds.

What are other forms of precipitation?

13. When bits of ice crystals form in clouds, they may fall to

    the ground as _____ .

14. Hailstones form inside the tall clouds of a _____
    and are usually the size of peas.

Critical Thinking

15. Describe how water changes form inside your house, like it
    does in the water cycle.

    _____

    _____

    _____

    _____

    _____

    _____

The Water Cycle

Match the correct word with its description.

a. condensation	**d.** melting	**g.** snow
b. evaporation	**e.** precipitation	**h.** water cycle
c. freezing	**f.** sleet	**i.** water vapor

1. _____ This is the condition that causes a liquid to change into a solid.

2. _____ This is the ongoing movement of water through many different processes and states.

3. _____ These are small drops of rain that freeze in the air before they hit the ground.

4. _____ This is a process where a liquid becomes a gas.

5. _____ These are ice crystals that form in clouds and fall to the ground.

6. _____ This is water that falls from clouds to Earth.

7. _____ This is the gas form of water.

8. _____ This is the process of a gas becoming a liquid.

9. _____ This is the process of a solid becoming a liquid.

The Water Cycle

Fill in the blanks.

cirrus	cumulus	stratus
clouds	evaporates	vapor
condenses	precipitation	water cycle

Water moves from the Earth to the atmosphere and

back again. This path is called the _____ .

Water changes to a gas, or _____ ,

from the surface of oceans, lakes, and other places.

Water _____ rises into the air and

cools. Then it _____ onto tiny particles

of dust and forms _____ .

There are three main types of clouds. Puffy white

clouds are called _____ clouds. Low,

layered clouds are called _____

clouds. Wispy clouds high in the sky are called

_____ clouds. Eventually, the water in

clouds falls back to Earth as _____ .

The different types of precipitation include rain, snow,

sleet, and hail.

Tracking the Weather

Use your textbook to help you fill in the blanks.

What are air masses and fronts?

1. A large region of air with nearly the same properties

   throughout is called a(n) _____ .

2. Many air masses form near the _____

   or near the _____ .

3. The boundary between two air masses is called

   a(n) _____ .

4. A warm air mass that pushes into a cold air mass is called

   a(n) _____ .

5. A cold air mass that pushes under a warm air mass is called

   a(n) _____ .

6. Two air masses that are not moving into each other form

   a(n) _____ .

What does a weather map show?

7. Weather maps show weather patterns. For example,
   lines of half circles or triangles show the locations of

   _____ .

© Macmillan/McGraw-Hill

8. Predicting weather conditions is called _____.

9. In the United States, fronts tend to move from

_____ to _____ .

10. Scientists use special instruments, such as _____

and _____ , to predict the weather.

What are the signs of severe weather?

11. The sound made when lightning quickly heats the air

around it is called _____ .

12. Thunderstorms can give rise to spinning winds that are

called _____ when they hit the ground.

13. Very wide storms that form over warm ocean water are

called _____ .

Critical Thinking

14. Why do you think the weather usually becomes cool
and clear after a severe thunderstorm?

_____

_____

_____

_____

_____

_____

_____

Name _____ Date _____

Tracking the Weather

Match the correct word with its description.

cold	hurricanes	tornadoes
forecast	mass	warm
front	thunderstorm	

1. The boundary between two air masses is called

   a(n) _____ .

2. Meteorologists study weather patterns and maps so that

   they can predict or _____ the weather.

3. If a front is fast-moving and bringing stormy weather,

   then it is a(n) _____ front.

4. A large region of air with nearly the same temperature

   and water vapor throughout is an air _____ .

5. A(n) _____ front is formed when warm
   air pushes into cold air and brings light, steady rain.

6. Although these very wide storms from over the ocean,

   _____ can also cause severe damage
   on land.

7. Heavy rain and lightning warn of an approaching

   _____ .

8. Rotating columns of air form _____
   that can reach speeds of 400 km (250 mi) per hour.

Tracking the Weather

Fill in the blanks.

air mass	equator	land	poles
cold front	front	oceans	warm front

The weather pattern on the ground depends on
what is happening in the air. The body of air that slowly
passes over a wide area of water or land is called a(n)
_____ . It takes on the characteristics
of the area in which it forms. For example, cold, dry, air
masses form over _____ , close to the
_____ . Warm, moist air masses form
over _____ , close to the _____ .

The place where two different air masses meet is
called a(n) _____ . A cold air mass
pushing under a warm air mass is called a(n)
_____ . A warm air mass pushing into
a cold air mass is called a(n) _____ .
To forecast the weather, scientists locate fronts
and track how they are moving.

Name _____ Date _____

Hurricane Season

Read the passage in your textbook. On the lines below, write the information that lets you know when and where hurricanes occur.

_____

_____

_____

Write About It
Fact and Opinion
1. What technologies help scientists study hurricanes?
2. What do you think would happen during a hurricane in your neighborhood?

_____

_____

_____

_____

_____

_____

_____

_____

Fill in the Fact and Opinion graphic organizer. Then, answer the question.

Fact	Opinion
Hurricanes usually happen in the _____ and northeast _____ oceans.	The National Hurricane Center in _____ , thinks there will be more hurricanes this year than last year.
There must be certain _____ for a hurricane to form.	The temperature of the ocean water isn't _____ enough for a hurricane to form until late June.
Hurricanes are storms that bring violent winds, large _____ , _____ , and lots of _____ .	Violent _____ may knock down trees, and large _____ may cause _____ .
Data about hurricanes comes from buoys, _____ , _____ , and supercomputers.	People tell what they _____ about hurricanes.

1. Why would a prediction be considered an opinion rather than a fact?

   _____

   _____

Climate

Use your textbook to help you fill in the blanks.

What is climate?

1. The pattern of seasonal weather that happens in an area

   year after year is called _____ .

2. Important factors that define climate are humidity, wind,

   _____ and _____ .

3. Temperate climates often have four _____ .

4. The types of _____ that farmers can
   grow depend on climate.

What determines climate?

5. The thin lines that run east and west across some

   maps are lines of _____ .

6. Latitude is a measure of how far a place is from the

   _____ and increases as you move
   north or south.

7. The temperature differences between low and high

   latitudes cause _____ .

8. Global winds are winds that move between the

   _____ and the _____ .

9. Warm air near the equator _____ and moves toward the poles; cold air near the poles

   _____ and moves toward the equator.

10. A directed flow of a gas or liquid is called a(n)

    _____ .

11. Water heats and cools more _____ than land does.

12. Climates near the ocean are cloudier and _____ than regions farther inland.

How do mountains affect climate?

13. The climate at the base of a mountain is always

    _____ than the climate at the peak.

14. As a(n) _____ travels over a mountain,

    it dries out. So the _____ on one side will be wetter than the climate on the other side.

Critical Thinking

15. What do you think the climate would be like if you lived at the base of a mountain near the ocean?

    _____

    _____

    _____

    _____

    _____

    _____

Climate

Match the correct letter with the description.

a. altitude	**d.** equator	**g.** mountain
b. climate	**e.** global winds	**h.** ocean current
c. climate region	**f.** latitude	

1. _____ the characteristic weather pattern of a region over the course of several years

2. _____ a measure of how far a place is from the equator

3. _____ an area with the same climate throughout

4. _____ a landform that can separate two different types of climates

5. _____ a measure of how high a place is above sea level

6. _____ the directed flow of water over long distances through the ocean

7. _____ winds that circulate the air between the equator and the poles

8. _____ where the latitude is set at zero degrees

Climate

Fill in the blanks.

altitude	latitude	precipitation
climate	land	temperatures
degrees	polar	tropical

The weather in a particular region can be averaged

over a long period of time. This is called the

_____ , and farmers depend on it to

grow their crops. Average yearly _____

and _____ define the climate of a region.

Areas on the equator have a(n) _____

of zero degrees and have _____ climates.

Latitude at the North and South Poles is 90

_____ , and they have _____

climates.

Air temperature decreases with _____ ,

so higher areas have cooler climates than lower areas.

Water warms and cools more slowly than _____

does. This is why areas near the ocean usually have milder

climates than inland areas.

Weather and Climate

Circle the letter of the best answer.

1. Which tool is used to measure air pressure?

   a. hygrometer

   b. anemometer

   c. barometer

   d. thermometer

2. The atmosphere is made mostly of nitrogen and

   a. oxygen.

   b. carbon dioxide.

   c. water vapor.

   d. hydrogen.

3. Which is the lowest layer of the atmosphere?

   a. stratosphere

   b. thermosphere

   c. ionosphere

   d. troposphere

4. Humidity is a measure of

   a. the weight of the air.

   b. the amount of water vapor in the air.

   c. precipitation.

   d. how hot or cold the air is.

5. A process during which a liquid changes into a gas is called

   a. condensation.

   b. freezing.

   c. evaporation.

   d. melting.

6. Dew forms on grass when water vapor from the air

   a. condenses.

   b. evaporates.

   c. melts.

   d. freezes.

7. Wispy clouds that form high in the sky are called

   a. cumulus clouds.

   b. stratus clouds.

   c. fog.

   d. cirrus clouds.

8. An air mass that forms over tropical ocean water will be

   a. warm and dry.

   b. cold and dry.

   c. warm and moist.

   d. cold and moist.

Circle the letter of the best answer.

9. A cold air mass pushing under a warm air mass is called a(n)

   a. warm front.

   b. cold front.

   c. stationary front.

   d. occluded front.

10. Fronts in the United States tend to move from

    a. west to east.

    b. east to west.

    c. north to south.

    d. south to north.

11. The widest type of storm is called a

    a. tornado.

    b. thunderstorm.

    c. winter storm.

    d. hurricane.

12. Global winds are caused by

    a. temperature differences between high and low latitudes.

    b. temperature differences between high and low altitudes.

    c. ocean currents.

    d. mountain ranges.

13. Which of the following will cause a climate to be cooler?

    a. lower altitude

    b. higher altitude

    c. lower latitude

    d. ocean current from the equator

14. Where does the latitude measure 0°?

    a. North Pole

    b. South Pole

    c. equator

    d. polar current

15. Which of the following is a measure of the weight of air pressing down on an area?

    a. air pressure

    b. temperature

    c. precipitation

    d. humidity

Name _____ Date _____

The Solar System and Beyond

Use the facts you have learned from the chapter to fill in the concept map.

The Universe

The universe is made up of 100 billion _____ ,

each with 200 billion _____ . Stars are glowing balls of gases.

The Solar System

The solar system includes an average star, _____ ,

eight orbiting _____ , many _____ ,

and smaller bodies, such as _____ and

_____ .

Earth

Earth and its Moon revolve around the _____

once each year. Earth is tilted on its _____ .

The Moon

As the Moon _____ around Earth

about once every _____ , it changes

_____ .

© Macmillan/McGraw-Hill

Earth and Sun

What causes day and night?

1. Earth completes one rotation on its _____

   every _____ hours.

2. As Earth _____ , the Sun appears

   to rise in the _____ and set in the

   _____ .

3. The stars, Moon, and planets appear to move across

   the sky each night because of Earth's _____ .

4. At dawn and dusk, shadows are _____ ,

   and at midday, they are _____ .

What causes seasons?

5. Each year, Earth completes one _____
   around the Sun.

6. In June, the North Pole is tilted _____
   the Sun, so sunlight hits the Northern Hemisphere at

   a(n) _____ angle.

7. In summer, light is more _____ than it
   it is in winter.

8. In December, the North Pole is tilted _____
   the Sun, so sunlight hits the Northern Hemisphere at

   a(n) _____ angle.

9. When it is winter in the Northern Hemisphere, it is

_____ in the Southern Hemisphere.

How does the Sun's apparent path change over the seasons?

10. In northern Alaska, summer nights are _____,

but during winter the Sun hardly _____ .

11. The Sun rises _____ in the sky in
summer than it does in winter.

12. Near the equator, the Sun's apparent path changes

_____ during the year than at higher
latitudes.

13. The Sun's path repeats itself every year, so it is

possible to predict the time the Sun will _____

and _____ .

Critical Thinking

14. What do you think would be different if Earth rotated
and revolved in the opposite direction? What would
stay the same?

_____

_____

_____

_____

_____

Earth and Sun

Match the correct word with its description. Write the letter of the word in the space provided.

a. apparent	**c.** hemisphere	**e.** revolution	**g.** seasons
b. axis	**d.** orbit	**f.** rotate	**h.** shadows

1. _____ the northern or southern half of Earth

2. _____ an invisible line that runs through the middle of an object

3. _____ what Earth does every 24 hours on its axis

4. _____ the path Earth takes around the Sun, or the path the Moon takes around Earth

5. _____ Earth's complete travel around the Sun

6. _____ what occurs because Earth orbits the Sun on a tilted axis

7. _____ the type of "motion" of the Sun as it rises in the east and sets in the west

8. _____ what changes during the day but always points away from the Sun

Earth and Sun

Fill in the blanks.

axis	higher	pattern	rotation
Earth	lower	poles	seasons
equator	path	revolution	Sun

Earth completes a spin every 24 hours. This

_____ causes day and night. It is day on

the part of Earth facing the _____ , and in

12 hours, it will be night.

Earth also completes a _____ around

the Sun each year. Because _____ is

revolving on a tilted _____ , there are

_____ . During the summer, the Sun rises

_____ in the sky and earlier in the day.

During the winter, the Sun is _____ in the

sky. Near the _____ , the temperature

and the Sun's apparent _____ change

very little. Near the _____ , the Sun has a

shorter apparent path but the same _____ .

Scientists use this information to predict the times the Sun

will rise and set.

© Macmillan/McGraw-Hill

Without the Sun

Write About It

Fictional Story Write your own story about what would happen, if sunlight could not reach Earth.

Getting Ideas

First
↓

Next
↓

Last

Planning and Organizing

A good story has characters, a setting, and a plot. Justin wrote three notes to plan his story. Write Character next to the note that mainly describes the character. Write Plot next to the note that mainly describes the plot. Write Setting next to the note that mainly describes the setting.

Note 1. _____   It is the year 5002, and total darkness has covered Planet Earth.

Note 2. _____   Professor Jamison is a scientist. Her specialty is the Sun.

Note 3. _____   Professor Jamison and her staff are trying to find out why Earth is suddenly in total darkness.

Revising and Proofreading

Here are some sentences that Justin wrote. He needs to include descriptive details. Choose a word from the box. Write it on the line.

black	brilliant	chilly	total

At first, there was a hint of darkness. The air became

_____ . Then, suddenly, there was

_____ darkness. The sky had been a

_____ blue. Now it was as _____

as the darkest ink.

Drafting

Begin your story. Start with an exciting sentence to get the reader interested.

_____

_____

Continue your story. Use a separate piece of paper. Include details that tell about the main character and the setting. Make sure your story tells what would happen if sunlight didn't reach Earth.

Now revise and proofread your writing. Ask yourself:

► Did I write an interesting beginning, middle, and end?

► Did I describe the characters and the setting?

► Did I correct all mistakes?

Earth and Moon

What is the Moon like?

1. Earth's closest neighbor in space is the _____ .

2. Moonlight is reflected light from the _____ .

3. The Moon has _____ similar to

those on Earth, but no _____

or _____ .

4. Temperatures on the Moon can be both _____
than any place on Earth.

5. The Moon's surface is covered by _____

made by _____ .

6. When meteoroids enter Earth's atmosphere, they become hot

and _____ before they hit Earth's surface.

What are the phases of the Moon?

7. The Moon orbits Earth once in just over _____
days.

8. At any given time, the Sun lights _____
of the Moon.

9. As the Moon orbits Earth, we see different parts of it lit

as it cycles through all of its _____ .

10. The Moon's _____ causes _____ ,
the daily rise and fall of the ocean's surface.

What is an eclipse?

11. A shadow cast by Earth or the Moon is a(n)

_____ .

12. Earth casts a shadow on the Moon during a(n)

_____ eclipse.

13. A lunar eclipse happens when Earth is directly

between _____ and _____ .

14. The Moon casts a shadow on Earth during a(n)

_____ eclipse.

15. A solar eclipse happens only when there is a(n)

_____ .

16. All of the Sun's light is blocked during

a(n) _____ .

Critical Thinking

17. Which do you think occurs more often, a partial solar
eclipse or a total solar eclipse? Explain your reasoning.

_____

_____

_____

_____

_____

Earth and Moon

Use the words from the word box to fill in the blanks.

crater	meteoroids	phases	tides
lunar eclipse	new Moon	solar eclipse	waning Moon

1. The Moon's gravity causes _____ .

2. The apparent shapes of the Moon in the sky are called

   its _____ .

3. The Moon casts a shadow on Earth during

   a(n) _____ .

4. A hollow pit in the ground is called a(n) _____ .

5. When the lighted side of the Moon faces away from

   Earth, it is called a(n) _____ .

6. Large rocks that fall from space are called _____ .

7. When less and less of the lighted side of
   the Moon becomes visible each night, it is

   a(n) _____ .

8. Earth casts a shadow on the Moon during

   a(n) _____ .

Name _____ Date _____

Earth and Moon

Fill in the blanks.

Earth	last-quarter	shadow
first-quarter	new Moon	three-fourths
full-Moon	one-fourth	

The Moon orbits Earth once every 29 days. When the

Moon and the Sun are on the same side of Earth, the

part of the Moon that is in _____ faces

Earth. This is the _____ phase. In

about a week, the Moon has completed _____

of its orbit. This is called the _____.

About a week later, Earth is between the Moon and

the Sun. This is called the _____ phase.

In another week, the Moon has completed _____

of its orbit and only half of the lighted side can be seen.

This is the _____ phase. During a solar

eclipse, the Moon casts a shadow on _____ .

During a lunar eclipse, Earth casts a shadow on the Moon.

The Solar System

What is the solar system?

1. Each planet revolves around the Sun in an orbit

   shaped like a(n) _____ , a slightly
   flattened circle.

2. Newton discovered that the balance between gravity

   and _____ keeps the planets in orbit.

3. In the 1500s, _____ proposed that the
   planets revolve around the Sun.

How do we learn about the solar system?

4. Telescopes make farway objects seem _____ .

5. In the 1960s, rockets from NASA took astronauts into

   _____ .

6. The United States worked with other countries to build

   the _____ , which can stay in space
   for a long time.

7. An unmanned spacecraft that carries data-recording

   equipment into space is called a(n) _____ .

What are the rocky planets?

8. Earth, Mars, _____ , and Venus are
   closest to the Sun and are called the rocky planets.

9. The atmosphere of Venus is made of _____ .

© Macmillan/McGraw-Hill

Name _____ Date _____

What are the other planets?

10. The four outer planets lie beyond _____ .

11. All of the outer planets are made mostly of hydrogen

and _____ .

12. The largest planet is _____ , and the

next-largest is _____ .

What else is in the solar system?

13. When comets get close to the Sun, they form

a(n) _____ .

14. Most asteroids lie in a belt between _____

and _____ .

Critical Thinking

15. What other planet would you like to live on? What do
you think would be the hardest thing to get used to?

_____

_____

_____

_____

_____

_____

_____

The Solar System

What am I?

Choose a word from the word box below that answers each question. Write the letter of the word in the space provided.

a. asteroid	**c.** gravity	**e.** meteorite	**g.** solar system
b. comet	**d.** meteor	**f.** planet	**h.** telescope

1. _____ I am the Sun and all of the objects that orbit it. What am I?

2. _____ I am one of the largest objects orbiting the Sun. What am I?

3. _____ I am an invisible pulling force that keeps the planets in orbit around the Sun. What am I?

4. _____ I can make distant objects appear to be closer. What am I?

5. _____ I am a chunk of ice mixed with rocks and dust. I travel around the Sun in a long, narrow orbit. What am I?

6. _____ I am made of chunks of rock or metal. I lie in a belt between Mars and Jupiter. What am I?

7. _____ I am a meteoroid that falls into Earth's atmosphere and burns up. What am I?

8. _____ I am a meteoroid that strikes Earth's surface. What am I?

The Solar System

Fill in the blanks.

comets	gases	Neptune	rocky
Earth	gas giants	planets	Venus
ellipses	Jupiter	rock	

The solar system consists of an average star, called

the Sun, and all of the objects that revolve around it.

These include eight _____ , many

moons, and several smaller bodies, such as asteroids

and _____ .

The _____ planets are Earth,

Mercury, _____ , and Mars. They are

closer to the Sun and are made mostly of _____ .

The _____ include _____ ,

Saturn, Uranus, and _____ . All of these

are made mostly of _____ . The orbits

of the planets are shaped like _____ .

Earth is the only planet in our solar system that has

what living things need to survive.

To the Moon!

How have scientists explored our solar system? What scientists learn about the Moon may help them explore planets and other solar system objects.

Write About It

Main Idea and Details Reread the introduction and the captions on the time line. Then write a paragraph that explains the main idea and details of this article. Be sure to include facts and examples in your paragraph.

Main Idea and Details

Fill in the Main Idea and Details Chart using information you find in the introduction and captions of the reading feature.

Main Idea	Details

Name _____ Date _____

Planning and Organizing

Answer these questions in more detail.

1. What was the first spacecraft to travel in space, and when was it launched?

   _____

2. What spacecraft was the first to land a person on the Moon, and when did this happen?

   _____

   _____

3. What was the last manned spacecraft to travel to the Moon, and when was it launched?

   _____

   _____

Drafting

Explain how people first learned about the far side of the Moon.

_____

_____

Are scientists still studying the Moon? Why?

_____

_____

_____

_____

Stars and Constellations

What are stars?

1. The closest star to Earth is _____ . It

   is _____ kilometers away from Earth.

2. The Sun is a(n) _____-sized star.

3. Red stars and orange stars are _____
   than the Sun, and blue stars and white stars

   are _____ .

4. The Sun will glow for _____ more
   years.

5. Scientists measure the distance of stars from Earth

   in _____ .

6. Throughout the universe, stars are found in large

   groups called _____ .

What are constellations?

7. Because of Earth's _____ , we see
   different constellations.

8. The constellations _____ to move
   across the sky throughout the year.

9. The night sky looks different in the _____

   than it does in the _____ .

© Macmillan/McGraw-Hill

10. Once there were no clocks and people used

constellations to tell _____ .

11. Long ago, farmers used constellations to tell them

when to plant or _____ crops.

12. Sailors used constellations to _____ .

What is the Sun like?

13. The Sun gives off energy in the form of _____

and _____ .

14. The Sun provides the energy needed by almost all of

Earth's _____ .

15. The Sun's energy powers _____ ,

_____ , and the water cycle.

Critical Thinking

16. Do you think it is possible to get a sunburn on a
cloudy day?

_____

_____

_____

_____

_____

_____

_____

Stars and Constellations

Use the words in the box to fill in the sentences and the crossword puzzle.

constellation	galaxy	Orion	Sun
Cross	light	star	

Across

3. A constellation visible from the Southern Hemisphere is called the Southern _____ .

5. A group of stars in the sky forms a(n) _____ .

7. The closest star to Earth is the _____ .

Down

1. The distance light travels in a year is called a(n) _____ -year.

2. A large group of stars is called a(n) _____ .

4. A constellation only visible in winter is _____ .

6. A hot, glowing sphere of gases is a(n) _____ .

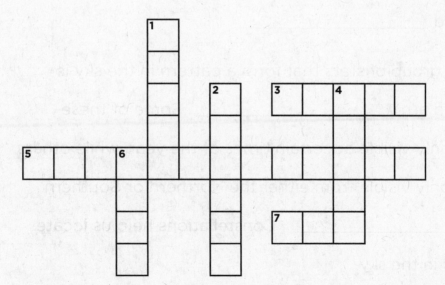

Name _____ Date _____

Stars and Constellations

Fill in the blanks.

galaxies	Hemispheres	reflect
blue or white	light-years	Sun
constellation	orange or red	

Each star in the sky is an enormous, hot, glowing

sphere of gases. Unlike planets that _____

light, stars emit their own light. The _____

is the closest star to Earth. It is average in temperature,

so it is yellow. Hotter stars are _____ ,

and cooler stars are _____ . The

distance of a star from Earth is measured in

_____ . Stars are found in large groups

called _____ .

A group of stars that form a pattern in the sky is

called a(n) _____ . Some of these

are only visible at certain times of the year, while others

are only visible from either the Northern or Southern

_____ . Constellations help us locate

stars in the sky.

© Macmillan/McGraw-Hill

The Solar System and Beyond

Circle the letter of the best answer.

1. When the North Pole is tilted toward the Sun, it is

 a. summer in the Northern Hemisphere.

 b. winter in the Northern Hemisphere.

 c. daytime in the Northern Hemisphere.

 d. nighttime in the Northern Hemisphere.

2. In June, the South Pole has

 a. almost 24 hours of daylight.

 b. almost 24 hours of darkness.

 c. 12 hours of daylight and 12 hours of darkness.

 d. 6 hours of daylight and 18 hours of darkness.

3. The Sun appears to move from east to west because, when looking down on the North Pole, Earth

 a. revolves counterclockwise.

 b. revolves clockwise.

 c. rotates counterclockwise.

 d. rotates clockwise.

4. When is a shadow the longest?

 a. noon in summer

 b. dawn in summer

 c. noon in winter

 d. dawn in winter

5. When Earth is between the Moon and Sun, we see a

 a. half Moon.

 b. full Moon.

 c. new Moon.

 d. gibbous Moon.

6. When half of the lighted side of the Moon is seen, we see a

 a. new Moon.

 b. quarter Moon.

 c. full Moon.

 d. crescent Moon.

7. A partial solar eclipse occurs during the

 a. full-Moon phase.

 b. new-Moon phase.

 c. gibbous-Moon phase.

 d. half-Moon phase.

© Macmillan/McGraw-Hill

Circle the letter of the best answer.

8. Which of the following have scientists discovered?

   a. Some planets orbiting distant stars have living things on them.

   b. As of yet, no planets orbiting distant stars have been discovered.

   c. Some distant stars have planets orbiting them.

   d. Some planets in our solar system have living things on them.

9. Which of the following is the hottest planet?

   a. Mercury

   b. Venus

   c. Earth

   d. Mars

10. Which of the following is a chunk of ice mixed with rocks and dust?

    a. comet

    b. asteroid

    c. meteoroid

    d. meteorite

11. One light-year is about

    a. one trillion km.

    b. ten billion km.

    c. ten trillion km.

    d. one billion km.

12. Which planet rotates on its side?

    a. Saturn

    b. Uranus

    c. Mars

    d. Neptune

13. Besides Earth, which other planet has ice caps?

    a. Venus

    b. Mars

    c. Mercury

    d. Uranus

14. Which star is closest to Earth?

    a. Sirius

    b. Proxima Centauri

    c. Ross 154

    d. Sun

Mr. Mix-It

by Nicole Iorio

from *Time for Kids*

Read the Unit Literature feature in your textbook.

Write About It

Response to Literature What type of job would you like to have when you grow up? What skills does it require? Write a paragraph about your plans.

_____

_____

_____

_____

_____

_____

_____

_____

_____

_____

_____

_____

Name _____ Date _____

Properties of Matter

Complete the concept map with the information you learned about conserving Earth's resources. Below each type of property in the concept map, fill in details or terms that relate to the type of property.

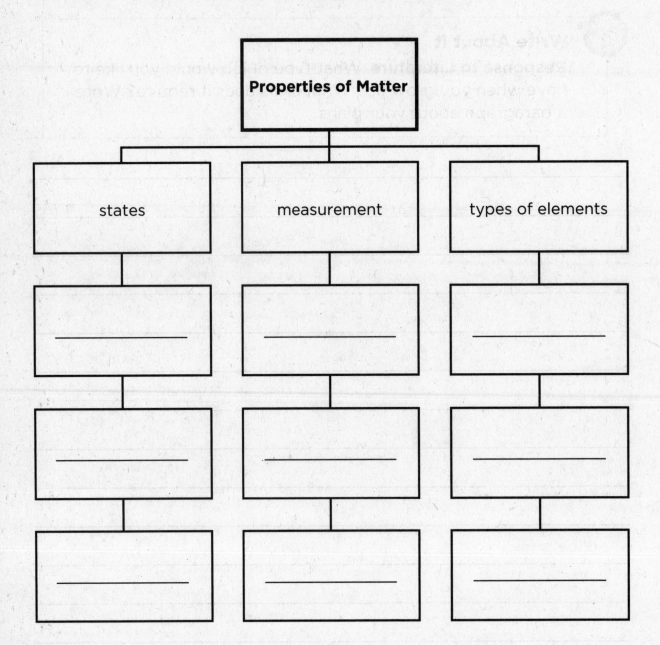

Describing Matter

Use your textbook to help you fill in the blanks.

What is matter?

1. Anything that has ___mass___ and takes up

   ___space___ is matter.

2. The amount of matter in an object is the

   object's ___mass___ .

3. Matter takes up space, so matter has ___volume___ .

4. Matter has many characteristics, or ___property___,
   that can be used to describe it.

5. To measure mass scientists use a tool called a

   ___balance___ units.

6. The ability of a material to ___dissolves___ in liquids
   is another property of some types of matter.

7. Properties that can not be seen can still be

   ___measured___ .

8. Wood floats on water because wood has the property

   of ___buoyancy___ .

What are the states of matter?

9. The different forms that matter can take are called

   _____ of matter.

10. The state of matter that has a definite shape and a definite

    volume is the ___solid___ state.

11. The _particles_ in a solid are packed tightly, often in a regular pattern.

12. When matter in the _liquid_ state is moved from one container to another, it keeps the same volume but takes the shape of its container.

13. Oxygen is a(n) _gas_ because it takes the shape and volume of the container in which it is placed.

14. Particles in a(n) _gas_ move about freely.

What happens to the matter we use?

15. When you use something again, you _recycle_ matter.

16. Matter can end up in _landfills_ or _oceans_ .

17. Matter can be _reuse_ , or made into something else.

18. Metal, paper, plastic, and _glass_ can be recycled.

Critical Thinking

19. Do you think you can describe an object without using its properties?

Yes, because a desk as matter in it.

© Macmillan/McGraw-Hill

Describing Matter

Choose a word from the word box below that correctly fills in the blank.

buoyancy	liquid	matter	solid
gas	mass	property	volume

1. A canoe floats because of _____ .

2. A large rock has more _____ than a small rock.

3. Because a table has mass and takes up space, it

   contains _____ .

4. Something is a _____ if it has a definite volume but takes the shape of its container.

5. Color is one _____ of an object.

6. If copper is moved from a jar to a dish, it keeps the same shape and volume because copper is

   a(n) _____ .

7. The amount of matter in a basketball is the

   _____ of the ball.

8. Oxygen in a car tire takes the shape and volume of the tire

   because oxygen is a(n) _____ .

Describing Matter

Fill in the blanks.

buoyancy	liquid	properties	volume
gas	mass	solid	

Matter can exist as three different states. If matter completely fills and takes the shape of its container, it is a(n) _____ . If matter always has the same shape and volume, even in different shaped containers, it is a(n) _____ . If a(n) _____ is poured from a long glass to a wide jar, it takes the shape of the jar but keeps the same volume.

Matter has characteristics, called _____ . One property is _____ , which describes the amount of matter something has. Another property is _____ , which describes how much space something takes up. Matter has the property of _____ , which allows some objects to float on a liquid.

Juggling Matter

> **Write About It**
> **Descriptive Writing** Choose three or four items to describe.
> For example, you might choose a child's toy, your pet's toy,
> and a backpack. Write a paragraph describing them. Include
> the properties that make these objects useful to you.

Getting Ideas

Make a list of three or four items that have something in
common. (For example, you might list the items in your
desk.) Then after each item, write words that you think
describe it.

1. _____

2. _____

3. _____

Planning and Organizing

Make sure the details you include appeal to at least one
of the five senses. Here are three sentences Valerie wrote.
Write "sight," "hearing," "smell," "taste," or "touch" next
to each sentence.

1. _____ My shiny silver calculator is very thin.

2. _____ I keep my pens and pencils in a soft case.

3. _____ There is a peppermint in my pocket.

Drafting

Write a sentence to begin your description. Tell what you are writing about and why you are writing.

_____

Now write your description on a separate piece of paper. Begin with the sentence you wrote above. Then describe each of the items. Include details that appeal to the senses, so that readers can picture the items. Tell how they are alike and different. At the end, tell how these items are part of a group.

Revising and Proofreading

Here is part of the description that Valerie wrote. Help her improve it by choosing a descriptive word from the box.

clattering	five	large	wooden

I have an old _____ desk at home

that holds some of my most treasured items. It has

_____ _____ drawers.

When I open a drawer, there is a _____

sound as the items move around.

Now revise and proofread your writing. Ask yourself:

▶ Did I use describing words to tell about the items?

▶ Did I compare and contrast them?

▶ Did I correct all mistakes?

Measurement

Use your textbook to help you fill in the blanks.

How do we measure matter?

1. An inch, a mile, a pound, and a gallon are standard

    _____ of measurement.

2. A system of measurement that is based on units of 10 is

    the _____ .

3. An object's _____ is the number of units
    that fit across.

4. Any measurement made in square units, such as
    square centimeters (cm^2), is a measurement

    of _____ .

What is density?

5. The comparison of an object's mass to its volume

    describes _____ .

6. To find the density of an object _____
    its mass by its volume.

7. An object floats when its density is _____
    than the density of the liquid or gas into which it is placed.

8. Heated air becomes _____ dense; cooler,
    denser air forces it upward.

9. The density of water is _____ while the density of cork is 0.24 g/cm³.

10. The density of an object affects its _____ .

What is weight?

11. The measure of the pull of gravity from a planet on the mass

of an object describes an object's _____ .

12. Ounces and pounds are the _____ units

for weight, and the _____ is the metric
unit for weight.

13. An object's _____ changes with gravity,

but its _____ stays the same.

Critical Thinking

14. Do you think that if a marshmallow and a marble are
the same size, they would have the same mass, density,
buoyancy, or volume?

_____

_____

_____

_____

Measurement

Match the correct letter with the description.

a. area	**d.** gravity	**g.** newton
b. balance	**e.** length	**h.** volume
c. density	**f.** metric system	**i.** weight

1. _____ a system of measurement based on the number 10

2. _____ the number of units across an object

3. _____ the effect of gravity on the mass of an object

4. _____ the force that pulls any two objects toward each other

5. _____ the mass of an object divided by its volume

6. _____ the metric unit of weight

7. _____ the number of square units on the surface of something

8. _____ the amount of space that an object takes up

9. _____ the tool used to measure mass

Name _Diya Sehgal_ Date _____

Measurement

Fill in the blanks.

balance	height	metric system	weight
gram	length	newton	
gravity	mass	volume	

Measurement is a way of using numbers to compare objects. The amount of matter in an object describes its _____ . Mass is measured by using a(n) _____ . The unit in the _____ that describes mass is the _____ . A measurement of the effect of the force of _____ on the mass of an object is the _____ of the object. The metric unit of weight is the _____ .

When the _____ of a box-shaped object is multiplied by its width and _____ , its _____ is found. The area of a flat surface can be calculated by multiplying its length by its width.

Classifying Matter

Use your textbook to help you fill in the blanks.

What are elements?

1. A material that contains only one type of matter is

   a(n) _____ .

2. An element _____ be broken down into a
   simpler form.

3. Elements are "the _____ blocks of matter."

4. An element is made of small particles, called

   _____ , each of which contains all of the
   properties of that element.

5. Elements are classified as _____ ,

   _____ , and _____ .

6. An element that conducts electricity and can be hammered

   into a shape is a(n) _____ .

7. Elements that have some, but not all, of the properties of

   metals are called _____ .

8. Nitrogen is a(n) _____ because it does
   not contain any of the properties of a metal.

How are elements organized?

9. The scientist _____ organized the
   elements according to their properties.

10. The table that shows the elements arranged according to

their properties is the _____ .

11. Rows in the periodic table are called _____ .

How do scientists use the periodic table?

12. Elements in the same _____ , such as
potassium and sodium, have similar properties.

13. Scientists use the periodic table to predict how an

element will behave or _____ .

14. Hydrogen reacts with, and forms, many _____ .

15. Iron and two other metals in its row are _____ .

16. Elements in the fluorine column react with elements in

_____ .

Critical Thinking

17. Do you think that new elements can still be added to
the periodic table of elements?

_____

_____

_____

_____

_____

Classifying Matter

Fill in the blank with the correct term. Then circle the term in the puzzle.

1. a material that contains only one type of matter:

   Element

2. a row in the periodic table: Period

3. a shiny element that conducts electricity: metal

4. an element with some, but not all, properties of metal:

   metalloid

5. elements are organized in a chart: Periodic table

6. an element with no properties of a metal: nonmetal

7. the small particles that make up an element:

   atims

```
E L B A T C I D O I R E P
P L A T E M N O N C A E E
M E T A L L O I D J O U X
E E A R E M D R O G P M E
T R X L M G N E F A U X T
A U O G E O U P S M O T A
L H H O N A T I M S R J L
Q V E R T N S D L M G U G
```

© Macmillan/McGraw-Hill

Name _____ Date _____

Classifying Matter

Fill in the blanks.

conduct	metals	periodic table
element	nonmetals	properties
metalloids	period	

The thousands of materials that we use every day are made from just over 100 different types of matter. Each type of matter is called a(n) _____ . Elements are classified in a chart called the _____ .

The periodic table organizes elements according to their _____ . Each row is called a(n) _____ . Elements in the same column have similar properties. Elements are also grouped in the periodic table as metals, _____ , and nonmetals. Metals are found on the left side of the periodic table. _____ are found on the right side. Metalloids have some but not all properties of _____ . Elements that are nonmetals have none of the properties of metals.

Meet Sisir Mondal

Read the passage in your textbook. Then use the Compare and Contrast graphic organizer to help you answer the questions.

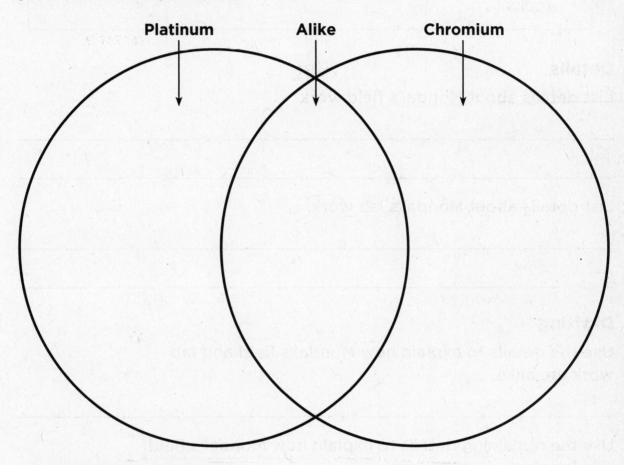

Platinum   Alike   Chromium

Compare and Contrast

1. How are platinum and chromium alike?

_____

2. How are platinum and chromium different?

_____

_____

Name _____ Date _____

Write About It

Classify Read the article again. What does Sisir look for in the rocks he studies? How do you think Sisir classifies the rocks?

Details

List details about Mondal's field work.

_____

_____

List details about Mondal's lab work.

_____

_____

Drafting

Use the details to explain how Mondal's field and lab work are alike.

_____

Use the remaining details to explain how Mondal's field and lab work are different.

_____

_____

_____

_____

Properties of Matter

Circle the letter of the best answer.

1. The smallest particle of gold cannot be broken down into simpler substances. Gold is a(n)

   a. atom.

   b. element.

   c. nonmetal.

   d. noble gas.

2. What property is being measured when the unit of measurement is cm^2?

   a. area

   b. length

   c. mass

   d. volume

3. A book keeps its shape and volume when it is moved from one bag to another. A book is a(n)

   a. gas.

   b. liquid.

   c. metal.

   d. solid.

4. What type of element is shiny and can be drawn into a wire?

   a. a metal

   b. a metalloid

   c. a noble gas

   d. a nonmetal

5. The noble gases in column 18 of the periodic table are all

   a. magnetic.

   b. nonmetals.

   c. metals.

   d. metalloids.

6. Which item makes up aluminum foil?

   a. a gas

   b. a metalloid

   c. a specific type of atom

   d. different kinds of atoms

7. Measurements of mass and volume can be used to find

   a. area.

   b. density.

   c. gravity.

   d. volume.

Circle the letter of the best answer.

8. The force that explains why wood floats on water is

   a. buoyancy.

   b. volume.

   c. gravity.

   d. mass.

9. When helium is moved from a tank to a balloon, its shape and volume change. Helium is a

   a. gas.

   b. liquid.

   c. metalloid.

   d. solid.

10. Gravity is stronger on Earth than it is on the Moon. On the Moon, an object's

   a. mass is greater than on Earth.

   b. mass is less than on Earth.

   c. weight is greater than on Earth.

   d. weight is less than on Earth.

11. Which measurement should you make in order to find the amount of carpet you need to order to cover a floor?

   a. area

   b. length

   c. density

   d. volume

12. Which of the following is based on units of ten?

   a. density

   b. gravity

   c. the metric system

   d. the periodic table

13. All matter has

   a. mass and area.

   b. mass and volume.

   c. state and area.

   d. weight and volume.

14. When a certain state of bromine is moved from a bottle to a jar, its volume stays the same but its shape takes the shape of the jar. Bromine in this state is a(n)

   a. gas.

   b. liquid.

   c. metal.

   d. solid.

© Macmillan/McGraw-Hill

How Matter Changes

Use your textbook to help you fill in the blanks.

Combinations of Matter

Mixtures

are made up of two
or more types of

_____.

Each type keeps its own

_____.

Two types of mixtures are

_____

and

_____.

Compounds

are made up of two or
more types of

_____.

This new substance has its own

_____.

Two types of compounds are

_____

and

_____.

Some mixtures can be
separated on the basis of

_____
properties.

**Mixtures can
be separated
using certain
techniques.**

Evaporating and
then recondensing
a liquid is called

_____.

Filtration separates liquids from

_____.

_____
turns a liquid into a gas.

Name _____ Date _____

How Matter Can Change

Use your textbook to help you fill in the blanks.

What are physical changes?

1. A change that begins and ends with the same kind of

   matter is a _____ .

2. A physical change can be caused by cutting, crushing,

   tearing, bending or _____ matter.

3. When the surface of a lake freezes and becomes

   _____ , the water beneath the ice

   is still a _____ .

4. Some signs of a physical change are changes in size,

   _____ , or _____ .

How does matter change state?

5. When matter changes from one form to another, such as
   from a solid to a liquid, it goes through a change

   of _____ .

6. A change of state is caused by _____ .

7. The Sun's energy can change a liquid to a gas during a

   process called _____ .

What are chemical changes?

8. A change that produces new matter is a(n)

   _____ .

© Macmillan/McGraw-Hill

9. A chemical change is also known as a chemical

_____ .

10. A chemical change either _____

or _____ energy.

11. Energy in a chemical change can be in the form of heat,

_____ , or _____ .

12. Some of the characteristics of a chemical change are the

release of a(n) _____ or a change in

_____ or smell.

What are other real-world changes?

13. Reshaping dough is a(n) _____ ,
but baking it is a chemical change.

Critical Thinking

14. Do you think you would identify a change in texture as a
chemical change or a physical change?

_____

_____

_____

_____

_____

_____

Name _____ Date _____

How Matter Can Change

Choose a word from the word box below that correctly fills in the blank.

change of state	compound	physical	tarnish
chemical	evaporation	rust	

1. When an ice cube melts into liquid water, it goes through

   a(n) _____ .

2. Folding, cutting, chopping, and crushing are examples of

   a(n) _____ change.

3. A chemical reaction between iron and oxygen
   found in the air and in water forms a new substance

   called _____ .

4. A change in matter that results in new matter is a(n)

   _____ change.

5. Energy from the Sun can change a liquid into a gas. The

   process is called _____ .

6. The result of two or more elements chemically combining

   to form a new substance is a(n) _____ .

7. When silver metal in a spoon reacts to sulfur in the air, a

   new substance, called _____ , is formed.

Name _____ Date _____

How Matter Can Change

Fill in the blanks.

change of state	physical	substance
chemical	properties	sulfur
evaporation	rust	tarnish
iron	solid	

Matter changes every day. A _____

change is the simplest type of change, because the

physical _____ change but the type of

matter remains the same. If water becomes a gas through

_____ , or if it freezes and becomes a

_____ , that is a_____ .

During a _____ change, a new

_____ forms. For example, when

oxygen and _____ combine,

_____ is formed. When silver combines

with _____ in the air, _____

is formed. These are two common chemical changes.

Name _____ Date _____

Lady Liberty

Read the passage in your textbook.

Write About It
Sequence

1. Make a chart showing how the color of the Statue of Liberty has changed over time.

2. Use your chart to write a summary of those changes.

Reread the text. Find any words you think are time-order words and then write them on the lines below.

1. _____

2. _____

3. _____

4. _____

5. _____

6. _____

Can you think of any statues or monuments in your city that have changed color over time?

_____

_____

_____

Sequence

Use the sequence-of-events chart to illustrate how the color of the Statue of Liberty has changed.

First

First, the statue was the color of a shiny new _____ .

Then, the statue turned from _____ to dark

_____ .

↓

Next

The dark brown color was caused by a _____ called

oxidation. During oxidation, _____ in the air

combined with the _____ of the statue to form

_____ .

↓

Last

Over time, _____ and _____ reacted

with the copper oxide. A new _____ , copper hydroxide,

formed. Copper hydroxide gave the statue its _____ .

Summary

Use your chart to write a paragraph on a separate sheet of paper that summarizes the Statue of Liberty's changes in appearance.

Mixtures

Use your textbook to help you fill in the blanks.

What is a mixture?

1. A physical combination of two or more kinds of matter is

   called a(n) _____ .

2. Each kind of matter in a mixture keeps its own

   _____ properties.

3. Many everyday products, including the _____

   you eat and the _____ you wear,
   are mixtures.

4. A mixture that is blended completely is called a(n)

   _____ .

5. Salt water is a good _____ of electricity.

6. A solution might have _____ that the
   original materials did not have.

7. A mixture of two or more metals or a metal and a nonmetal

   is called a(n) _____ .

How can you separate the parts of a mixture?

8. The materials in a mixture can be separated by using their

   _____ .

9. Solids are often separated from a liquid by using

   a(n) _____ .

© Macmillan/McGraw-Hill

10. Using a filter to separate a mixture is

called _____ .

11. A mixture containing certain metals can be separated

using a(n) _____ .

How can you separate the parts of a solution?

12. Solids can be separated from a solution by using

_____ ; the liquid part of the solution
is lost to the air.

13. Liquid water can be collected from a solution of

salt water by using _____ .

Critical Thinking

14. How are mixtures useful in your everyday life?

_____

_____

_____

_____

_____

_____

_____

_____

_____

Name _____ Date _____

Mixtures

What am I?

Choose a word from the word box below that answers each question. Write the corresponding letter on the line.

a. alloy	**c.** filter	**e.** mixture
b. distillation	**d.** filtration	**f.** solution

1. _____ I am a combination of two or more types of matter in which each type of matter keeps its own chemical properties. What am I?

2. _____ I am a mixture in which substances are completely blended throughout the mixture. What am I?

3. _____ I am a solution of two or more metals. I can also be a solution of a metal and a nonmetal. What am I?

4. _____ I am a tool that can be used to separate the parts of some mixtures. I am often used to separate a solid from a liquid. What am I?

5. _____ I am a process used to separate a liquid solution by heating it and collecting the gas. What am I?

6. _____ I am the process used to separate parts of a mixture by using a mesh or screen. What am I?

Mixtures

Fill in the blanks.

alloy	evaporation	solution
chemical properties	filter	
distillation	physical properties	

People use mixtures every day. Each type of matter

in a mixture keeps its own _____ .

Parts of a mixture can be separated by using their

_____ . Two tools that can be

used to separate a mixture are a magnet and a(n)

_____ . Filtration is used to separate

a solid from a liquid.

Two substances can also be blended completely to form

a(n) _____ . When metals are blended

together, they form a(n) _____ , which may

be harder and stronger than the original materials. Solutions

can be separated through _____ , used

to remove the solids, and _____ , used to

remove the liquids. Mixtures and solutions can be separated

because their parts have kept their own physical properties.

Compounds

Use your textbook to help you fill in the blanks.

What are compounds?

1. Not all combinations of _____ can be separated physically.

2. Two or more elements can be combined chemically to

   form a substance called a(n) _____ .

3. A compound from sodium and chlorine that you might find

   on the dinner table is _____ .

4. No amount of crushing can separate the _____ in table salt.

5. A common compound formed from hydrogen and

   oxygen is _____ .

6. A common compound formed from iron and oxygen

   is _____ .

7. _____ , hydrogen, and oxygen combine
   to form sugar.

8. Silicon and oxygen combine to form _____ ,
   a hard mineral.

What are acids and bases?

9. Acids and _____ are compounds that

   _____ easily with other substances.

10. Some substances, like lemons and oranges, contain

    _____ acids.

11. Scientists use _____ paper to tell whether a substance is an acid or a base.

12. A(n) _____ is a substance that turns red litmus paper blue.

13. A(n) _____ is a substance that turns blue litmus paper red.

14. When an acid and a base are _____ , they react chemically.

15. Combining an acid and a base will form a _____

and _____ .

16. Water _____ turn litmus paper blue or red.

Critical Thinking

17. How do we use acids and bases around the house?

_____

_____

_____

_____

_____

_____

_____

_____

_____

Compounds

Match the correct word with the description.

acid	compound	neutral	physically
base	litmus paper	opposites	water

1. When two or more elements are chemically combined, a(n)

   _____ is formed.

2. Scientists identify acids and bases by using

   _____ .

3. Lemon juice can be used in food and to clean because it is

   a weak _____ .

4. Not every substance is an acid or a base. Some substances,

   such as water, are _____ .

5. Acids and bases have properties that are _____
   of one another.

6. A compound made from hydrogen and oxygen

   is _____ .

7. Soap feels slippery and can be tested with red litmus paper

   because it is a(n) _____ .

8. If elements have been chemically combined, they cannot

   be separated _____ .

Compounds

Fill in the blanks.

> chemical reaction   hydrogen   oxygen
>
> chemicals   litmus paper   substance

Matter is made of elements that have been joined

together. When two or more elements are joined

through a(n) _____ , they form a new

_____ with new properties. Water

is an example of a compound made of two gases:

_____ and _____ .

When these two gases are joined, they produce a new

substance—liquid water.

Acids and bases are substances that are present in many

everyday products. Acids, bases, and neutral substances

can be identified using _____ . Strong

acids and bases are harmful _____ . Weak

acids and bases are used in food and soap.

How Matter Changes

Circle the letter of the best answer.

1. If a piece of paper changes in size or shape it most likely indicates that
   a. a chemical change has taken place.
   b. a physical change has taken place.
   c. a change of state has taken place.
   d. evaporation has taken place.

2. What type of change occurs when liquid water is cooled to make ice cubes?
   a. change of place
   b. change of shape
   c. change of state
   d. change in use

3. Which describes evaporation?
   a. gas changed into a liquid
   b. liquid changed into a gas
   c. liquid changed into a solid
   d. gas changed into a solid

4. The type of change that produces a new substance is a
   a. physical change.
   b. chemical change.
   c. change of state.
   d. change in place.

5. Iron and oxygen react chemically to form
   a. tarnish.
   b. dust.
   c. rust.
   d. heat.

6. Silver metal chemically reacts with sulfur in the air to form
   a. tarnish.
   b. dust.
   c. rust.
   d. heat.

© Macmillan/McGraw-Hill

Circle the letter of the best answer.

7. What is the name of a mixture of two or more metals?

    a. alloy

    b. rust

    c. tarnish

    d. base

8. What process involves heating a solution until a gas forms, collecting the gas, and then cooling the gas back into a pure liquid?

    a. evaporation

    b. filtration

    c. solution

    d. distillation

9. A mesh or screen that allows for the separation of a solid from a liquid is a(n)

    a. evaporate.

    b. filter.

    c. distillate.

    d. solution.

10. The name of the process used to separate a solid from a liquid, such as separating spaghetti from water, is

    a. evaporation.

    b. distillation.

    c. filtration.

    d. solution.

11. A combination of two or more types of matter that keep their original chemical properties is called a(n)

    a. mixture.

    b. solution.

    c. alloy.

    d. chemical reaction.

12. What two types of substances have opposite properties and form a neutral solution when they are combined?

    a. acids and gases

    b. acids and bases

    c. bases and gases

    d. bases and water

© Macmillan/McGraw-Hill

Name _____ Date _____

Magnetic Migration

From *Ranger Rick*

Read the Unit Literature feature in your textbook.

Write About It

Response to Literature Have you taken a trip to a different place? Where did you go? How did you get there? Write about a trip you have taken. Be sure to include how you figured out the directions.

_____

_____

_____

_____

_____

_____

_____

_____

_____

_____

_____

_____

_____

Forces

Fill in the blanks in this concept map by using the information you have learned about forces.

Motion		Forces

Motion

is a change in the

_____

of an object
and has speed,

_____ ,

and sometimes

_____ .

Forces

can start or

_____

motion, or they can

_____

the direction of
motion. Forces include

_____

and

_____ .

Motion and Forces

Use your textbook to help you fill in the blanks.

What is motion?

1. Motion occurs when an object changes its location or

   its _____ .

2. Words like left, _____ , _____ ,

   below, east, and _____ give clues
   about position.

3. When you describe how far you walk by what
   you see along the way, those things are your

   _____ of reference.

4. The word used to describe how far apart two points or

   places are is _____ .

5. The _____ of anything is the distance
   it has moved in a certain period of time.

6. Speed is a change in _____ over time.

7. A moving object's _____ describes its

   _____ and the _____
   it moved.

How do forces affect motion?

8. You must apply a(n) _____ to put an

   object in motion or _____ an object
   from moving.

© Macmillan/McGraw-Hill

9. Acceleration is any change in the _____

or direction of a(n) _____ object.

10. Any object is in a state of _____
until a force moves it or stops it.

11. Objects in motion can be slowed down or stopped by

_____ once they touch each other.

What is gravity?

12. The force that pulls objects together is

called _____ .

13. The pull of gravity between two objects is affected by

the amount of _____ in each object

and the _____ between the objects.

Critical Thinking

14. Do you think you need a lot of force to change an
object's state of inertia?

_____

_____

_____

_____

_____

_____

Motion and Forces

Match the correct word with the description.

a. acceleration	**c.** friction	**e.** inertia	**g.** time
b. force	**d.** gravity	**f.** speed	**h.** velocity

1. _____ the distance that an object moves in a certain amount of time

2. _____ can be used to determine speed if distance is known

3. _____ a change in the speed or direction of a moving object

4. _____ a push or a pull that can move a still object or stop a moving object

5. _____ the tendency of an object in motion to stay in motion, or an object at rest to stay at rest

6. _____ the type of force when the surfaces of two objects touch

7. _____ the force that pulls objects together

8. _____ the speed and direction in which an object is moving

Motion and Forces

Fill in the blanks.

force	inertia	speed
friction	position	stop
gravity	rest	velocity

Objects do not move until a force acts on them.

They have _____ , the property that

keeps an object at _____ or in motion.

A(n) _____ is required to _____

a moving object, or to put a stopped object in motion.

Two of the forces that affect motion are _____

and _____ .

A moving object changes its _____ ,

or location, in relation to surrounding objects. An

object can move quickly or slowly, but the average

_____ of an object is equal to the

distance traveled divided by the time spent moving.

The speed and direction in which an object is moving is

called its _____ . A change in velocity

is called acceleration.

Name _____ Date _____

Changing Motion

Use your textbook to help you fill in the blanks.

How do forces affect motion?

1. When you put a backpack on a desk, _____ pulls the backpack toward Earth.

2. The backpack on a desk does not move. The desk

   _____ up on the backpack with a force

   that is _____ to the pull of gravity.

3. Forces that cancel each other out when acting on a

   single object are _____ .

4. Forces that are not equal to each other are called

   _____ .

5. Unbalanced forces cause a change in _____ ;

   the _____ force determines the direction
   of the motion.

6. Newtons are a measure of force in _____
   units.

How do forces affect acceleration?

7. The acceleration of an object is affected when

   the size of the _____ acting on
   it changes; acceleration is also affected by the

   _____ of the object.

8. The greater the mass the greater the force needed to

overcome _____ .

9. If the same force is used on two objects, the object

with less mass accelerates _____ than
the object with more mass.

How does friction affect motion?

10. Friction is a(n) _____ that works

_____ motion.

11. The amount of friction depends on the _____
involved.

12. To reduce friction on the moving parts of a bicycle, you

can use _____ .

Critical Thinking

13. How do you think firemen use forces and changing
motion to slide down a pole?

_____

_____

_____

_____

_____

_____

_____

Changing Motion

Choose a word from the word box below that correctly
fills in the blank.

balanced	friction	newtons	smooth
force	mass	rough	unbalanced

1. A(n) _____ surface creates a lot
   of friction.

2. Forces that are equal in size and opposite in direction

   are called _____ forces.

3. Any push or pull is called a(n) _____ .

4. Force is measured in units called _____ .

5. A force that works against motion is _____ .

6. Forces that are not equal in size and are not opposite

   in direction are called _____ forces.

7. More force is needed to move an object with a

   large _____ .

8. A(n) _____ surface does not create
   much friction.

Changing Motion

Fill in the blanks.

balanced forces	motion	speed
direction	newtons	unbalanced forces
friction	rough	

An object in motion is affected by different forces.

These forces, measured in units called _____,

can keep an object in _____ or cause

it to stop moving.

If forces are equal in size and move in opposite

directions, they are called _____ and do

not cause a change in motion. If they are not equal in

size and are opposite in _____ , they

are called _____ . They can affect an

object's direction, _____ , or both. The

motion created by an unbalanced force is also affected

by the type of surface under an object. A(n)

_____ surface produces more

_____ than a smooth surface. A

smooth surface will increase motion and a rough surface

will slow it down.

Name _____ Date _____

Wheels in Motion

Write About It

Explanatory Writing Research how the brakes on a bike work. Write a description that explains how friction helps the bike stop moving.

Getting Ideas

First find out how brakes work. Then fill out the chart below. Write the steps in the process.

First

⬇

Next

⬇

Last

Planning and Organizing

Jada wrote three steps for her explanation. Put the steps in the correct order. Write 1, 2, and 3.

_____ This causes friction between the brake and the wheel. It makes the bike come to a complete stop.

_____ The brake cable attaches to the back wheel. When you squeeze the brake cable, it tightens the brake.

_____ The handlebar brake lever is on the front handlebar. It is attached to the brake cable.

Revising and Proofreading

Here are some sentences that Jada wrote. Combine each pair. Use the word in parentheses. Put a comma before the word.

1. Don't press the brakes quickly. You might topple over the front wheels. (or)

   _____

   _____

2. Friction is created between the brake and the wheel. This causes the bike to come to a stop. (and)

   _____

   _____

Drafting

Begin your explanation. Write a topic sentence. Tell what your explanation is about.

_____

_____

Now write your explanation. Use a separate piece of paper. Start with the sentence you wrote above. Write easy-to-follow details to tell how the brakes on a bicycle work.

Now revise and proofread your writing. Ask yourself:

► Did I describe how the brakes on a bicycle work?

► Did I give clear, easy-to-follow details?

► Did I correct all mistakes?

Name _____ Date _____

Work and Energy

Use your textbook to help you fill in the blanks.

What is work?

1. Work is done when a force moves a(n) _____

   a certain _____ .

2. Any time you _____ or _____
   to move an object, you do work.

3. Energy is the ability to do _____ .

4. Stored _____ with the future ability to

   do work is called _____ .

5. A moving object has the energy of motion,

   called _____ .

What are some forms of energy?

6. Different forms of energy are chemical energy,

   _____ energy, light energy, mechanical

   energy, thermal energy, and _____ .

How can energy change?

7. Energy can be _____ from one object
   to another.

8. When kinetic energy is transferred from one
   marble to another, the second marble picks up

   _____ .

© Macmillan/McGraw-Hill

9. When energy is _____ , it changes form.

10. Examples of transformation of energy include the

change of electrical energy to _____
energy in a blender and the change of electrical energy to

light and _____ energy in a light bulb.

Critical Thinking

11. Put the forms of energy in order according to which form of energy people depended on first. Explain your reasoning.

_____

_____

_____

_____

_____

_____

_____

_____

_____

_____

_____

Work and Energy

Use the clues below to help you find the words hidden in the puzzle.

1. When force is used
to move an object,

_____

is done.

2. The ability to do work is

_____ .

3. Stored energy is called

_____

energy.

4. Matter in motion has

_____

energy.

5. A(n) _____
occurs when energy changes
from one form to another.

6. Chemical energy is converted
to electrical energy in a(n)

_____ .

7. Work is done when
a force moves a(n)

_____ .

```
K S X A R W A T O P I S A H
O B Q I P F N B N O P B E M
C B C E N E R G Y T D L R E
K A J N E G K I N E T I C W
T T I E I S F V J N V U H N
S T E E C D X G K T R D A C
C E F E L T Z R P I B G E L
T R A N S F O R M A T I O N
P Y U E N W X Y N L B N Y C
```

© Macmillan/McGraw-Hill

Work and Energy

Fill in the blanks.

electrical	motion	transferred
energy	potential energy	transformed
kinetic energy	stored	work
light	thermal	

Work is being done when a force moves an object.

There is no _____ being done if an

object is not moved. The ability to do work is called

_____ . An object with _____

energy is able to do work. This energy is also known as

_____ . When an object begins moving,

the potential energy changes to _____ .

Kinetic energy is the energy of _____ .

When energy is _____ , it passes

from one object to another. When energy is

_____ , it changes form. Every day,

different forms of energy, such as _____

energy, _____ energy, _____

energy, and mechanical energy are used. Nuclear energy

comes from splitting the tiniest particles of matter.

Name _____ Date _____

Hybrid Power

Read the passage in your textbook. As you read, write down the topic sentence of each paragraph.

Topic sentence:

1. _____

2. _____

   _____

3. _____

4. _____

5. _____

 Write About It

Summarize Read the article again. How do hybrid cars work? How do hybrid cars help the environment?

1. How do hybrid cars help people?

   _____

2. How do hybrid cars work?

   _____

   _____

3. How does that help the environment?

   _____

Reread the article. As you read, record the details that support each of the sentences in the Summarize graphic organizer.

Main Idea	Details
	The gasoline we use is made from oil, a nonrenewable _____ .
	In a traditional car, the _____ engine runs all the time.
	Hybrid cars use two power sources: _____ and _____ .
Hybrid cars that use _____ and _____ energy can lessen our _____ on gasoline and reduce _____ pollution.	A hybrid car uses less _____ and switches to a(n) _____ motor powered by _____ when the car slows down or comes to a stop.
	The batteries _____ when the car comes to a stop.
	The gasoline engine in a(n) _____ car is small and _____ efficient.

Simple Machines

Use your textbook to help you fill in the blanks.

What are simple machines?

1. A device that uses force to do work is

   a(n) _____ .

2. The force you apply to do work with a lever is called

   the _____ .

3. A machine that has only a few parts is called

   a(n) _____ .

4. A lever makes work easier by reducing the amount

   of _____ needed to move a load.

5. A lever works in two ways: A small force can be

   applied over a long _____ , or a large

   force can be applied over a(n) _____
   distance.

What are two other simple machines?

6. A fixed pulley can be used to change the _____

   of a force; a movable pulley can change the _____
   of a force needed to move a load.

7. A "wheel and axle" is a simple machine made up of a

   wheel and a bar _____ to the wheel.

What are inclined planes?

8. An inclined plane is a simple machine with a

_____ , slanted surface.

9. An inclined plane twisted into a(n) _____
is a screw.

10. With an inclined plane, you need only a _____

effort force, but you have to move the load _____ .

How do simple machines work together?

11. A _____ is made up of a combination
of simple machines.

12. How much work a machine produces compared
to how much work is applied is the machine's

_____ .

Critical Thinking

13. Which do you think are more useful, simple machines
or compound machines? Why?

_____

_____

_____

_____

_____

_____

Simple Machines

What am I?

Choose a word from the word box below that answers each question.

a. compound machine	**e.** load
b. effort force	**f.** pulley
c. inclined plane	**g.** simple machine
d. lever	

1. _____ I am a machine that has only a few parts, and I am used to make work easier. What am I?

2. _____ I am a machine made up of a bar, the movable part, and a fulcrum, the fixed point. What am I?

3. _____ I am the object being moved by a machine. What am I?

4. _____ I am the amount of force needed to do work. What am I?

5. _____ I am a machine that uses a wheel to change the direction or amount of force needed to move an object. What am I?

6. _____ I am a machine made up of only one part—a flat, slanted surface. What am I?

7. _____ I am a combination of two or more simple machines. What am I?

Simple Machines

Fill in the blanks.

amount	efficiency	wheels
combination	parts	
compound machines	simple machines	

Machines help us by making work easier to do.

Machines change the _____ of effort

force needed to do work. There are two main types of

machines: _____ and compound

machines. Simple machines have only a few

_____ . There are six kinds of simple

machines: levers, pulleys, _____ and

axles, inclined planes, wedges, and screws.

Most machines are _____ made

from a(n) _____ of simple machines.

Some compound machines use gears to help the parts of the

simple machines work together. The _____

of a machine is the measure of how much work it does

compared to the amount of work it requires.

Forces

Circle the letter of the best answer.

1. A train is traveling west at 80 kilometers per hour. The "80 kilometers per hour" is the train's

   a. speed.

   b. velocity.

   c. acceleration.

   d. direction.

2. What tendency of objects is described in the following statement?
   A moving object will stay in motion until a force acts on it. An object at rest will stay at rest until a force acts on it.

   a. inertia

   b. acceleration

   c. speed

   d. velocity

3. The force that acts on the surfaces along which objects touch is

   a. inertia.

   b. velocity.

   c. gravity.

   d. friction.

4. What force between objects is affected by the amount of matter in the objects and the distance between the objects?

   a. inertia

   b. gravity

   c. friction

   d. speed

5. What type of force can NOT cause a change in motion?

   a. gravity

   b. friction

   c. unbalanced force

   d. balanced force

6. What type of force can cause a change in an object's direction, speed, or both?

   a. unbalanced force

   b. balanced force

   c. acceleration

   d. inertia

Circle the letter of the best answer.

7. Force is measured in units called

   a. newtons.

   b. kilometers per hour.

   c. centimeters.

   d. meters.

8. When a force is used to move an object a certain distance, what is being done?

   a. energy

   b. work

   c. inertia

   d. friction

9. Energy is the

   a. result of friction.

   b. motion of an object.

   c. ability to do work.

   d. result of inertia.

10. What is the scientific term for the energy of motion?

   a. inertia

   b. movement

   c. kinetic energy

   d. potential energy

11. The force applied to a simple machine is called the

   a. effort force.

   b. work force.

   c. strong force.

   d. distance force.

12. Any kind of object that is being moved by a simple machine is called a

   a. lever.

   b. load.

   c. fulcrum.

   d. pulley.

13. A simple machine made of a wheel and a cord is

   a. a lever.

   b. a pulley.

   c. a wedge.

   d. a screw.

14. A compound machine is made of

   a. several different parts.

   b. a combination of two or more simple machines.

   c. gears that do work.

   d. electrical parts.

© Macmillan/McGraw-Hill

Name _____ Date _____

Energy

Fill out the following concept map with information that you have learned from the chapter.

Heat is the flow of energy from

a(n) _____ object to

a(n) _____ object.
Heat is transferred by conduction,

_____ , and
radiation.

Sound is produced when

_____ particles
create waves. Sound has

_____ , pitch,
amplitude, and

_____ .

Kinds of Energy

Electricity can be static or

_____ . Static
electricity is a build-up of

_____ charges in
a material. Current electricity
is a constant flow of electricity

through a(n) _____ .

Light travels in _____ .

Light can _____
off of a surface or

_____ when it
moves from one material to
another.

Magnetism is found in magnets

and _____ . Magnets

have a(n) _____
that attracts iron, nickel, and
cobalt. Electric current creates
electromagnets, which can be
turned off with a(n)

_____ .

© Macmillan/McGraw-Hill

Heat

Use your textbook to help you fill in the blanks.

What is heat?

1. Heat is the flow of _____ from one object to another.

2. Heat flows from a(n) _____ object to a(n) _____ object.

3. When you toast bread, you also heat the _____ around the bread. Touch the toast and _____ moves to your _____ .

4. Hot particles _____ as they _____ thermal energy.

5. When something is heated, its _____ changes.

6. A tool that is used to measure temperature is a(n) _____ .

7. Fahrenheit and Celsius are two _____ used to measure _____ .

How does heat travel?

8. Heat is transferred by _____ , _____ , and radiation.

9. The transfer of heat through space is called _____ .

10. Matter that transfers heat easily is a(n)

_____ .

11. Matter that does not transfer heat easily is a(n)

_____ .

How does heat change matter?

12. Particles of matter are always _____ .

13. Adding heat makes particles of matter move

_____ and spread farther

_____ .

14. Heat can cause matter to _____ or

_____ .

Critical Thinking

15. What materials do you think would make the best
potholders? Hint: What do bakers use to take pizza,
bread, and cookies out of the oven?

_____

_____

_____

_____

_____

_____

_____

© Macmillan/McGraw-Hill

Heat

Match the correct word with its description.

a. conduction	**c.** convection	**e.** insulator	**g.** radiation
b. conductor	**d.** heat	**f.** melting	**h.** temperature

1. _____ a material that transfers heat easily

2. _____ the transfer of heat through space

3. _____ the flow of thermal energy from one object to another

4. _____ the transfer of energy that occurs between two objects that are touching

5. _____ a measure of the average kinetic energy of the particles in a substance

6. _____ the process of heat transfer in liquids and gases

7. _____ a material that does not easily conduct heat

8. _____ a change of state caused by heating

Heat

Fill in the blanks.

Celsius	gas	particles
conduction	increases	temperature
conductors	liquid	thermometer

Heat flows from a warmer object to a cooler object. When heat is added to an object, the object's _____ rises. A(n) _____ uses a Fahrenheit or a(n) _____ scale to show temperature changes in an object.

The thermal energy of particles of matter _____ when heat is added and _____ move faster. Heat is transferred from one material to another by _____ , convection, and radiation. Materials called _____ transfer heat easily. Materials called insulators do not.

Heat can change a solid to a(n) _____ or a liquid to a(n) _____ . Heat changes matter in many ways!

Sound

Use your textbook to help you fill in the blanks.

What is sound?

1. A vibration is a back-and-forth _____

   that produces _____ .

2. As the bell of an alarm clock vibrates, it causes nearby

   _____ particles to move.

3. A wave that transfers sound through matter is a(n)

   _____ .

4. Sound waves transfer energy from particle to particle
   and move away from a vibrating object

   in _____ directions.

How does sound travel?

5. Sound waves travel through _____ ,

   liquids, and _____ .

6. A sound that bounces off a surface is a reflected

   sound, or a(n) _____ .

7. Sound waves travel through different _____

   at different _____ .

8. Sound travels slowest in a gas. It travels more quickly

   through a(n) _____ , and travels most

   quickly through a(n) _____ .

9. Dolphins use echoes to navigate and find _____ .

10. You hear sound when sound waves in the air make tiny

_____ in your ears vibrate.

How do sounds differ?

11. The distance from the top of one sound wave to the

top of the next wave is a(n) _____ .
The number of wavelengths that pass a point in one

second is called _____ .

12. The highness or lowness of a sound is its _____ .

13. _____ affects the volume, or
loudness, of sound; it is related to the amount of

_____ in a sound wave.

What is sonar?

14. We use sound waves in a(n) _____
device to detect objects under water.

15. Sailors use sonar to measure how _____
the water is.

Critical Thinking

16. Do you think there is sound in outer space?

_____

_____

_____

_____

Sound

What am I?

**Choose a word from the word box below that answers
each question.**

a. amplitude	**d.** pitch	**g.** volume
b. echo	**e.** sound wave	**h.** wavelength
c. frequency	**f.** vibration	

1. _____ I am the back-and-forth motion of an object.
   What am I?

2. _____ I am a wave that transfers sound through
   matter. What am I?

3. _____ I am a sound that has been reflected off a
   surface. What am I?

4. _____ I am the distance from the top of one sound
   wave to the top of the next sound wave.
   What am I?

5. _____ I am the number of wavelengths that pass a
   point in one second. What am I?

6. _____ I am the highness or lowness of a sound.
   What am I?

7. _____ I am the amount of energy in a sound wave.
   What am I?

8. _____ I am the loudness or softness of a sound.
   What am I?

Sound

Fill in the blanks.

bones	directions	particles	vibrations
brain	echo	speeds	waves

Have you ever wondered how you hear sound?

Sound waves travel through your outer ear, where

_____ send the vibrations to the inner

ear. A nerve carries sound messages to the _____ .

All sounds are produced by _____

and travel in _____ . When a bell rings,

sound waves move outward as _____

of air bump into one another. The waves move away

from the bell in all _____ . A sound wave

may bounce off a nearby surface as a reflected sound,

or _____ .

Sound waves travel through various materials at

different _____ . Sound travels slowest

through a gas and fastest through a solid.

The Voice in the Well

Write About It

Personal Narrative Have you heard an echo? What made the sound? Write a personal narrative about your experience.

Getting Ideas

Use the chart below to plan your personal narrative.

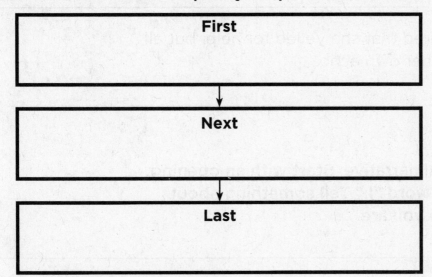

First

Next

Last

Planning and Organizing

Nikki wrote three sentences. Put the sentences in time order. Write 1 by the event that happened first. Write 2 by the event that happened next. Write 3 by the event that happened last.

_____ Then I yelled really loudly.

_____ I heard my voice echo through the canyon.

_____ I climbed down the steps and walked to the middle of the canyon.

Revising and Proofreading

Here are some sentences that Nikki wrote. She forgot to use the first-person pronoun "I" to write about her experience. Rewrite each sentence in the first person.

1. She liked hearing her voice bounce off the canyon walls.

_____

2. She could hear the echo of her voice.

_____

3. She was so scared that she yelled for help, but all she heard was her own echo.

_____

Drafting

Begin your personal narrative. Start with an opening sentence using the word "I." Tell something about yourself. Tell where you are.

_____

_____

Now write your personal narrative. Use a separate piece of paper. Start with the opening sentence you wrote above. Write the events in time order. Use time-order words. Include a beginning, a middle, and an end.

Revising and Proofreading

Now revise and proofread your writing. Ask yourself:

▶ Did I use the first-person pronoun "I" to tell the story?

▶ Did I use time-order words?

▶ Did I correct all mistakes?

Light

Use your textbook to help you fill in the blanks.

What is light?

1. Light is a form of _____ we detect with our eyes.

2. A tool used to separate white light into different colors

is a(n) _____ .

3. The colors that make up white light are called

the _____ .

How does light travel?

4. Light rays _____ as they pass from one material to another.

5. Light travels more slowly through _____ materials.

6. A lens is a tool used to _____ , or bend, light.

7. A lens that bends light outward, making objects look

smaller, is called a(n) _____ lens. A lens that bends light toward its center, making objects

look bigger, is called a(n) _____ lens.

8. The lens of an eye focuses images on the _____ .

The optic nerve sends images to the _____ .

© Macmillan/McGraw-Hill

What is reflection?

9. Reflection occurs when light strikes and then

   _____ a surface. Smooth, shiny

   surfaces, such as _____ , reflect the
   most light.

10. The law of reflection involves two light rays: the

    _____ ray and the outgoing ray. The

    angles of both rays are _____ .

What can light pass through?

11. Opaque material blocks light, _____
    material allows light to pass through, and translucent
    material allows some light to pass through but

    _____ it in different directions.

12. For privacy, people use _____
    materials.

Critical Thinking

13. Why do you think you should avoid wearing black
    clothing on a hot, sunny day?

    _____

    _____

    _____

    _____

    _____

Light

Match the correct word with its description.

---

a. electromagnetic	**d.** reflection	**g.** transparent
b. opaque	**e.** refraction	**h.** visible spectrum
c. prism	**f.** translucent	

---

1. _____ a tool used to separate white light into all of its colors

2. _____ the spectrum that encompasses all of the wavelengths of light

3. _____ the bending of light rays as they pass through different materials

4. _____ all of the colors we see that make up white light

5. _____ a material through which light cannot pass

6. _____ a material through which light can pass

7. _____ a material through which light can pass but will be scattered in different directions

8. _____ the property of light in which light rays strike a mirror and bounce off

Light

Fill in the blanks.

blocked	reflection	transparent
concave	refraction	two
mirrors	translucent	

Light has certain properties. It passes through some

materials and is _____ by others.

Opaque materials block light, _____

materials let some light pass through, and _____

materials allow all light to pass through.

The process in which light waves bend as they pass

from one transparent material to another is called

_____ . Lenses refract light in different

ways. Two kinds of lenses are _____

and convex lenses.

Light can also bounce off an object. This is called

_____ . Smooth, shiny surfaces, such as

_____ , reflect the most light. Reflection

involves _____ light rays: an incoming ray

and an outgoing ray. The angles of both rays are equal.

© Macmillan/McGraw-Hill

Electricity

Use your textbook to help you fill in the blanks.

What is electrical charge?

1. All matter is made up of tiny particles called

   _____ .

2. Atoms contain particles with both _____

   and _____ electrical charges.

3. We can show a positive charge as a(n) _____

   and a negative charge as a(n) _____ .

4. Opposite charges _____ each other,

   and like charges _____ each other.

5. The _____ of electrical charges on a

   material is called _____ electricity.

How do charges move?

6. Lightning is a(n) _____ of static

   electricity during a _____ .

7. The path along which electric current flows is called

   a(n) _____ .

8. A continuous flow of electricity through a circuit is

   called _____ electricity. An unbroken

   circuit is called a(n) _____.

9. A circuit with a gap is a(n) _____
   circuit; a circuit can be opened and closed with a(n)

   _____ .

What are series and parallel circuits?

10. Electric current flows along a single path in a

    _____ circuit.

11. A parallel circuit connects each load to the power

    source by different paths called _____.

How can you use electricity safely?

12. A fuse box and a circuit breaker are two safety devices

    that _____ a circuit when the current

    of electricity is dangerously _____.

Critical Thinking

13. Why do you think there's only one shock after you rub
    your socks (while your feet are in them) on the carpet
    and then touch the doorknob, but when you rub a
    balloon on the curtains or carpet and place it on the
    wall, it stays there for a long period of time?

    _____

    _____

    _____

Electricity

What am I?

Choose a word from the word box below that answers each question.

a. circuit	**d.** parallel circuit	**g.** switch
b. current electricity	**e.** series circuit	
c. discharge	**f.** static electricity	

1. _____ I am a buildup of negative charges on an object. What am I?

2. _____ I am the fast movement of charge from one object to another. What am I?

3. _____ I am made up of parts that work together to form a complete path, allowing electricity to flow. What am I?

4. _____ I am the result of the continuous flow of electricity through a circuit. What am I?

5. _____ I control the flow of electric current through a circuit. What am I?

6. _____ I am a circuit in which one loop of wire connects all of my parts. I will not work if one of my parts is removed. What am I?

7. _____ I am a circuit in which each part is connected to the power source through a separate path. I will continue to work if one of my parts is removed. What am I?

Electricity

Fill in the blanks.

atom	moves	series circuit
circuit	parallel circuit	short circuit
equal	rubbed	static electricity

Matter is made up of negatively and positively

charged particles. A(n) _____ usually

has a(n) _____ number of opposite

charges. When two materials, such as a balloon and

cloth, are _____ together, a buildup of

_____ forms. Discharge occurs when a

charge _____ quickly from one object

to another.

Current electricity flows through a path called a(n)

_____ . A(n) _____

contains parts connected with one loop of wire. A(n)

_____ contains objects connected to

a power source with separate paths. Frayed wires in

a(n) _____ can cause a fire. Fuses and

circuit breakers prevent the excessive flow of current.

© Macmillan/McGraw-Hill

Magnetism and Electricity

Use your textbook to help you fill in the blanks.

What is a magnet?

1. A magnet is an object or a material that can

    _____ metals containing _____ ,
    nickel, or cobalt.

2. Magnets can attract and _____
    each other.

3. All magnets have a north pole and a(n) _____
    pole.

4. The north poles and south poles of two magnets will

    _____ each other.

5. The attraction of magnets is _____
    when magnets are close together.

What are magnetic fields?

6. A magnetic field is the area of magnetic _____
    around a magnet.

What is an electromagnet?

7. An electromagnet is a coil of wire wrapped around a

    _____ core; current moving through

    the wire can be turned off with a(n) _____ .

8. Electromagnets are often used to power electric

    _____ .

What is a generator?

9. Almost all electricity is produced by _____ .

10. Generators produce back-and-forth current called

    _____ current. Batteries produce

    current in one direction, called _____
    current.

11. Air, steam, and _____ turn turbines.

How does electricity get to your home?

12. Electricity from a power plant is carried to a series

    of _____ .

13. Each transformer in the series increases or decreases

    the current's _____ until it is at the
    right level to enter your home.

Critical Thinking

14. Do you think there is a difference between the simple
    motor used in the electric cars your parents may drive
    and the model cars you and your friends may race?

    _____

    _____

    _____

    _____

    _____

Magnetism and Electricity

Match the correct word with the description.

a. attract	**d.** generator	**g.** pole
b. electromagnet	**e.** magnetic field	**h.** repel
c. electric motor	**f.** motor	

1. _____ how the north pole of one magnet and the south pole of another magnet react to each other

2. _____ how the north poles or south poles of two magnets react to each other

3. _____ where the force of a magnet is strongest

4. _____ the area of magnetic force around a magnet

5. _____ a magnet created by an electric current

6. _____ a device that changes electrical energy into mechanical energy

7. _____ a device that changes mechanical energy into electrical energy

8. _____ made of a power source, a magnet, a rotating loop of wire, and a motor shaft

© Macmillan/McGraw-Hill

Name _____ Date _____

Magnetism and Electricity

Fill in the blanks.

| distance | electromagnet | magnetic field | motors |
| electrical | generators | mechanical | transformers |

Magnets attract or repel each other. The

_____ between two magnets

determines the strength of their magnetic force. The

magnetic force around a magnet is its _____ .

Electric current creates a(n) _____

that can be turned off with a switch. Electromagnets

power _____ that change electrical

energy into _____ energy.

The electricity we use is produced at power plants by

_____ that change mechanical energy

into _____ energy. The voltage of

electricity is adjusted in _____ . Then

it is sent to our homes.

Motors at Work

Refrigerators, vacuum cleaners, hair dryers, and fans have one thing in common: they all have a motor. You can use those motors today because of people such as Joseph Henry and Michael Faraday. In 1831, these two scientists discovered how to use electromagnets to turn electrical energy into motion.

A few years later, Thomas Davenport, a blacksmith in Vermont, learned about electromagnets and built the first simple motor. He used the device to separate iron from iron ore.

It wasn't long before people started inventing new devices that used motors. Washing machines, invented in the early 1900s, use a motor to turn and wash your clothes. Another motor in a washing machine turns the water faucet on and off. Some of the first automobiles ran on electrical energy. Today many new cars use electric motors in addition to gasoline engines. Motors are useful for a lot of things! Can you think of any other machines that use electric motors?

Write About It

Problem and Solution What problem did Thomas Davenport solve with his motor? Write about a problem you have had, such as a messy room. How did an electric motor help you solve the problem?

_____

_____

_____

Problem and Solution

Use the Problem and Solution graphic organizer below to record the problem you need to solve and your plan for solving it.

Problem

↓

Steps to Solution

↓

Solution

Energy

Circle the letter of the best answer.

1. A form of energy that always moves from a warmer object to a cooler object is
   a. electricity.
   b. sound.
   c. light.
   d. thermal.

2. Which process transfers heat through liquids or gases?
   a. conduction
   b. convection
   c. radiation
   d. acceleration

3. Which of these has magnetic properties and can be controlled by a switch?
   a. an electric current
   b. an electromagnet
   c. a magnetic field
   d. a magnetic force

4. What is produced when energy causes particles to vibrate in waves?
   a. light
   b. sound
   c. electricity
   d. heat

5. Which of these terms describes the highness or lowness of a sound?
   a. wavelength
   b. pitch
   c. echo
   d. frequency

6. A device that indicates the direction of Earth's North Pole is a
   a. generator.
   b. motor.
   c. compass.
   d. sonar.

Circle the letter of the best answer.

7. The entire range of waves that make up light is the
   a. white light.
   b. electromagnetic spectrum.
   c. visible spectrum.
   d. law of reflection.

8. Which term describes how light bends as it passes from one transparent material into another?
   a. reflection
   b. refraction
   c. vibration
   d. convection

9. What type of material allows some light to pass through?
   a. convex
   b. transparent
   c. translucent
   d. concave

10. Which of these is produced by the buildup of negative charges on a material?
    a. conduction
    b. magnetism
    c. static electricity
    d. current electricity

11. Which of these is made up of parts that work together to allow electricity to flow?
    a. circuit
    b. insulator
    c. current
    d. electromagnet

12. The movement of static electricity is called a
    a. circuit.
    b. dlscharge.
    c. wavelength.
    d. frequency.

13. The region of magnetic force around a magnet is called a
    a. magnetic field.
    b. visible spectrum.
    c. compass.
    d. generator.